Use I

Advisors:
Robert Gentleman · Kurt Hornik · Giovanni Parmigiani

Use R!

Christian Kleiber · Achim Zeileis

Applied Econometrics with R

 Springer

Christian Kleiber
Universität Basel
WWZ, Department of Statistics and Econometrics
Petersgraben 51
CH-4051 Basel
Switzerland
Christian.Kleiber@unibas.ch

Achim Zeileis
Wirtschaftsuniversität Wien
Department of Statistics and Mathematics
Augasse 2–6
A-1090 Wien
Austria
Achim.Zeileis@wu-wien.ac.at

Series Editors
Robert Gentleman
Program in Computational Biology
Division of Public Health Sciences
Fred Hutchinson Cancer Research Center
1100 Fairview Avenue N., M2-B876
PO Box 19024, Seattle, Washington 98102-1024
USA

Kurt Hornik
Department of Statistics and Mathematics
Wirtschaftsuniversität Wien
Augasse 2–6
A-1090 Wien
Austria

Giovanni Parmigiani
The Sidney Kimmel Comprehensive Cancer Center
at Johns Hopkins University
550 North Broadway
Baltimore, MD 21205-2011
USA

ISBN: 978-0-387-77316-2   e-ISBN: 978-0-387-77318-6
DOI: 10.1007/978-0-387-77318-6

Library of Congress Control Number: 2008934356

Printed on acid-free paper

springer.com

Preface

R is a language and environment for data analysis and graphics. It may be considered an implementation of S, an award-winning language initially developed at Bell Laboratories since the late 1970s. The R project was initiated by Robert Gentleman and Ross Ihaka at the University of Auckland, New Zealand, in the early 1990s, and has been developed by an international team since mid-1997.

Historically, econometricians have favored other computing environments, some of which have fallen by the wayside, and also a variety of packages with canned routines. We believe that R has great potential in econometrics, both for research and for teaching. There are at least three reasons for this: (1) R is mostly platform independent and runs on Microsoft Windows, the Mac family of operating systems, and various flavors of Unix/Linux, and also on some more exotic platforms. (2) R is free software that can be downloaded and installed at no cost from a family of mirror sites around the globe, the Comprehensive R Archive Network (CRAN); hence students can easily install it on their own machines. (3) R is open-source software, so that the full source code is available and can be inspected to understand what it really does, learn from it, and modify and extend it. We also like to think that platform independence and the open-source philosophy make R an ideal environment for reproducible econometric research.

This book provides an introduction to econometric computing with R; it is not an econometrics textbook. Preferably readers have taken an introductory econometrics course before but not necessarily one that makes heavy use of matrices. However, we do assume that readers are somewhat familiar with matrix notation, specifically matrix representations of regression models. Thus, we hope the book might be suitable as a "second book" for a course with sufficient emphasis on applications and practical issues at the intermediate or beginning graduate level. It is hoped that it will also be useful to professional economists and econometricians who wish to learn R. We cover linear regression models for cross-section and time series data as well as the common nonlinear models of microeconometrics, such as logit, probit, and tobit models, as well as regression models for count data. In addition, we provide a chapter on programming, including simulations, optimization, and an introduction to `Sweave()`—an environment that allows integration of text and code in a single document, thereby greatly facilitating reproducible research. (In fact, the entire book was written using `Sweave()` technology.)

We feel that students should be introduced to challenging data sets as early as possible. We therefore use a number of data sets from the data archives of leading applied econometrics journals such as the *Journal of Applied Econometrics* and the *Journal of Business & Economic Statistics*. Some of these have been used in recent textbooks, among them Baltagi (2002), Davidson and MacKinnon (2004), Greene (2003), Stock and Watson (2007), and Verbeek (2004). In addition, we provide all further data sets from Baltagi (2002), Franses (1998), Greene (2003), and Stock and Watson (2007), as well as selected data sets from additional sources, in an R package called **AER** that accompanies this book. It is available from the CRAN servers at `http://CRAN.R-project.org/` and also contains all the code used in the following chapters. These data sets are suitable for illustrating a wide variety of topics, among them wage equations, growth regressions, dynamic regressions and time series models, hedonic regressions, the demand for health care, or labor force participation, to mention a few.

In our view, applied econometrics suffers from an underuse of graphics— one of the strengths of the R system for statistical computing *and graphics*. Therefore, we decided to make liberal use of graphical displays throughout, some of which are perhaps not well known.

The publisher asked for a compact treatment; however, the fact that R has been mainly developed by statisticians forces us to briefly discuss a number of statistical concepts that are not widely used among econometricians, for historical reasons, including factors and generalized linear models, the latter in connection with microeconometrics. We also provide a chapter on R basics (notably data structures, graphics, and basic aspects of programming) to keep the book self-contained.

The production of the book

The entire book was typeset by the authors using LaTeX and R's `Sweave()` tools. Specifically, the final manuscript was compiled using R version 2.7.0, **AER** version 0.9-0, and the most current version (as of 2008-05-28) of all other CRAN packages that **AER** depends on (or suggests). The first author started under Microsoft Windows XP Pro, but thanks to a case of theft he switched to Mac OS X along the way. The second author used Debian GNU/Linux throughout. Thus, we can confidently assert that the book is fully reproducible, for the version given above, on the most important (single-user) platforms.

Settings and appearance

R is mainly run at its default settings; however, we found it convenient to employ a few minor modifications invoked by

```
R> options(prompt="R> ", digits=4, show.signif.stars=FALSE)
```

This replaces the standard R prompt > by the more evocative R>. For compactness, `digits = 4` reduces the number of digits shown when printing numbers from the default of 7. Note that this does not reduce the precision with which these numbers are internally processed and stored. In addition, R by default displays one to three stars to indicate the significance of p values in model summaries at conventional levels. This is disabled by setting `show.signif.stars = FALSE`.

Typographical conventions

We use a `typewriter` font for all code; additionally, function names are followed by parentheses, as in `plot()`, and class names (a concept that is explained in Chapters 1 and 2) are displayed as in "`lm`". Furthermore, boldface is used for package names, as in **AER**.

Acknowledgments

This book would not exist without R itself, and thus we thank the R Development Core Team for their continuing efforts to provide an outstanding piece of open-source software, as well as all the R users and developers supporting these efforts. In particular, we are indebted to all those R package authors whose packages we employ in the course of this book.

Several anonymous reviewers provided valuable feedback on earlier drafts. In addition, we are grateful to Rob J. Hyndman, Roger Koenker, and Jeffrey S. Racine for particularly detailed comments and helpful discussions. On the technical side, we are indebted to Torsten Hothorn and Uwe Ligges for advice on and infrastructure for automated production of the book. Regarding the accompanying package **AER**, we are grateful to Badi H. Baltagi, Philip Hans Franses, William H. Greene, James H. Stock, and Mark W. Watson for permitting us to include all the data sets from their textbooks (namely Baltagi 2002; Franses 1998; Greene 2003; Stock and Watson 2007). We also thank Inga Diedenhofen and Markus Hertrich for preparing some of these data in R format. Finally, we thank John Kimmel, our editor at Springer, for his patience and encouragement in guiding the preparation and production of this book. Needless to say, we are responsible for the remaining shortcomings.

May, 2008

Christian Kleiber, Basel
Achim Zeileis, Wien

Contents

1

Introduction

This brief chapter, apart from providing two introductory examples on fitting regression models, outlines some basic features of R, including its help facilities and the development model. For the interested reader, the final section briefly outlines the history of R.

1.1 An Introductory R Session

For a first impression of R's "look and feel", we provide an introductory R session in which we briefly analyze two data sets. This should serve as an illustration of how basic tasks can be performed and how the operations employed are generalized and modified for more advanced applications. We realize that not every detail will be fully transparent at this stage, but these examples should help to give a first impression of R's functionality and syntax. Explanations regarding all technical details are deferred to subsequent chapters, where more complete analyses are provided.

Example 1: The demand for economics journals

We begin with a small data set taken from Stock and Watson (2007) that provides information on the number of library subscriptions to economic journals in the United States of America in the year 2000. The data set, originally collected by Bergstrom (2001), is available in package **AER** under the name Journals. It can be loaded via

```
R> data("Journals", package = "AER")
```

The commands

```
R> dim(Journals)
```

```
[1] 180  10
```

```
R> names(Journals)
```

C. Kleiber, A. Zeileis, *Applied Econometrics with R*,
DOI: 10.1007/978-0-387-77318-6_1, © Springer Science+Business Media, LLC 2008

```
[1] "title"        "publisher"    "society"     "price"
[5] "pages"        "charpp"       "citations"   "foundingyear"
[9] "subs"         "field"
```

reveal that Journals is a data set with 180 observations (the journals) on 10 variables, including the number of library subscriptions (subs), the price, the number of citations, and a qualitative variable indicating whether the journal is published by a society or not.

Here, we are interested in the relation between the demand for economics journals and their price. A suitable measure of the price for scientific journals is the price per citation. A scatterplot (in logarithms), obtained via

```
R> plot(log(subs), * log(price/citations), data = Journals)
```

and given in Figure 1.1, clearly shows that the number of subscriptions is decreasing with price.

The corresponding linear regression model can be easily fitted by ordinary least squares (OLS) using the function lm() (for linear model) and the same syntax,

```
R> j_lm <- lm(log(subs), * log(price/citations), data = Journals)
R> abline(j_lm)
```

The abline() command adds the least-squares line to the existing scatterplot; see Figure 1.1.

A detailed summary of the fitted model j_lm can be obtained via

```
R> summary(j_lm)
```

```
Call:
lm(formula = log(subs) ~ log(price/citations),
   data = Journals)

Residuals:
    Min      1Q  Median      3Q     Max
-2.7248 -0.5361  0.0372  0.4662  1.8481

Coefficients:
                     Estimate Std. Error t value Pr(>|t|)
(Intercept)            4.7662     0.0559    85.2   <2e-16
log(price/citations)  -0.5331     0.0356   -15.0   <2e-16

Residual standard error: 0.75 on 178 degrees of freedom
Multiple R-squared: 0.557,          Adjusted R-squared: 0.555
F-statistic:  224 on 1 and 178 DF,  p-value: <2e-16
```

Specifically, this provides the usual summary of the coefficients (with estimates, standard errors, test statistics, and p values) as well as the associated R^2, along with further information. For the journals regression, the estimated

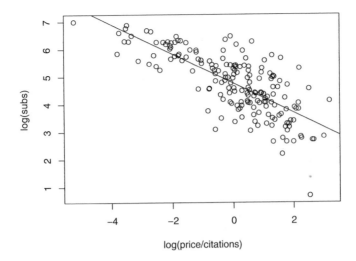

Fig. 1.1. Scatterplot of library subscription by price per citation (both in logs) with least-squares line.

elasticity of the demand with respect to the price per citation is -0.5331, which is significantly different from 0 at all conventional levels. The $R^2 = 0.557$ of the model is quite satisfactory for a cross-section regression.

A more detailed analysis with further information on the R commands employed is provided in Chapter 3.

Example 2: Determinants of wages

In the preceding example, we showed how to fit a simple linear regression model to get a flavor of R's look and feel. The commands for carrying out the analysis often read almost like talking plain English to the system. For performing more complex tasks, the commands become more technical as well—however, the basic ideas remain the same. Hence, all readers should be able to follow the analysis and recognize many of the structures from the previous example even if not every detail of the functions is explained here. Again, the purpose is to provide a motivating example illustrating how easily some more advanced tasks can be performed in R. More details, both on the commands and the data, are provided in subsequent chapters.

The application considered here is the estimation of a wage equation in semi-logarithmic form based on data taken from Berndt (1991). They represent a random subsample of cross-section data originating from the May 1985

Current Population Survey, comprising 533 observations. After loading the
data set CPS1985 from the package **AER**, we first rename it for convenience:

```
R> data("CPS1985", package = "AER")
R> cps <- CPS1985
```

For cps, a wage equation is estimated with log(wage) as the dependent variable and education and experience (both in number of years) as regressors.
For experience, a quadratic term is included as well. First, we estimate a
multiple linear regression model by OLS (again via lm()). Then, quantile regressions are fitted using the function rq() from the package **quantreg**. In
a sense, quantile regression is a refinement of the standard linear regression
model in that it provides a more complete view of the entire conditional distribution (by way of choosing selected quantiles), not just the conditional mean.
However, our main reason for selecting this technique is that it illustrates that
R's fitting functions for regression models typically possess virtually identical
syntax. In fact, in the case of quantile regression models, all we need to specify
in addition to the already familiar formula and data arguments is tau, the set
of quantiles that are to be modeled; in our case, this argument will be set to
0.2, 0.35, 0.5, 0.65, 0.8.

After loading the **quantreg** package, both models can thus be fitted as
easily as

```
R> library("quantreg")
R> cps_lm <- lm(log(wage) ~ experience + I(experience^2) +
+     education, data = cps)
R> cps_rq <- rq(log(wage) ~ experience + I(experience^2) +
+     education, data = cps, tau = seq(0.2, 0.8, by = 0.15))
```

These fitted models could now be assessed numerically, typically with a
summary() as the starting point, and we will do so in a more detailed analysis in Chapter 4. Here, we focus on graphical assessment of both models, in
particular concerning the relationship between wages and years of experience.
Therefore, we compute predictions from both models for a new data set cps2,
where education is held constant at its mean and experience varies over the
range of the original variable:

```
R> cps2 <- data.frame(education = mean(cps$education),
+     experience = min(cps$experience):max(cps$experience))
R> cps2 <- cbind(cps2, predict(cps_lm, newdata = cps2,
+     interval = "prediction"))
R> cps2 <- cbind(cps2,
+     predict(cps_rq, newdata = cps2, type = ""))
```

For both models, predictions are computed using the respective predict()
methods and binding the results as new columns to cps2. First, we visualize
the results of the quantile regressions in a scatterplot of log(wage) against
experience, adding the regression lines for all quantiles (at the mean level of
education):

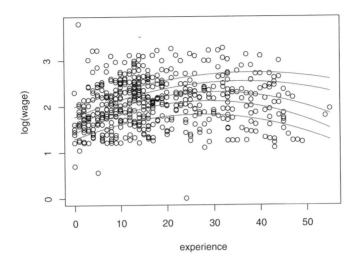

Fig. 1.2. Scatterplot of log-wage versus experience, with quantile regression fits for varying quantiles.

```
R> plot(log(wage) ~ experience, data = cps)
R> for(i in 6:10) lines(cps2[,i] ~ experience,
+     data = cps2, col = "red")
```

To keep the code compact, all regression lines are added in a `for()` loop. The resulting plot is displayed in Figure 1.2, showing that wages are highest for individuals with around 30 years of experience. The curvature of the regression lines is more marked at lower quantiles, whereas the relationship is much flatter for higher quantiles. This can also be seen in Figure 1.3, obtained via

```
R> plot(summary(cps_rq))
```

which depicts the changes in the regression coefficients over varying quantiles along with the least-squares estimates (both together with 90% confidence intervals). This shows that both `experience` coefficients are eventually decreasing in absolute size (note that the coefficient on the quadratic term is negative) with increasing quantile and that consequently the curve is flatter for higher quantiles. The intercept also increases, while the influence of `education` does not vary as much with the quantile level.

Although the size of the sample in this illustration is still quite modest by current standards, with 533 observations many observations are hidden due to overplotting in scatterplots such as Figure 1.2. To avoid this problem, and to further illustrate some of R's graphics facilities, kernel density estimates will

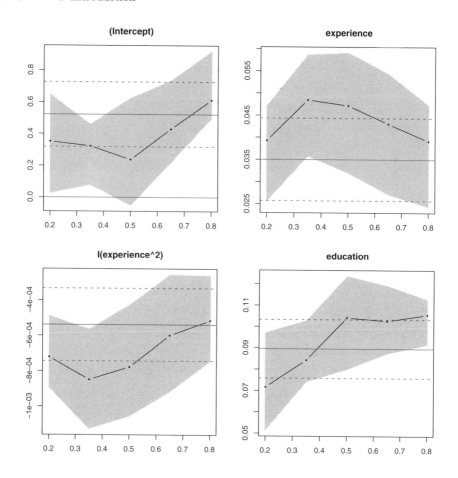

Fig. 1.3. Coefficients of quantile regression for varying quantiles, with confidence bands (gray) and least-squares estimate (red).

be used: high- versus low-density regions in the bivariate distribution can be identified by a bivariate kernel density estimate and brought out graphically in a so-called heatmap. In R, the bivariate kernel density estimate is provided by bkde2D() in the package **KernSmooth**:

```
R> library("KernSmooth")
R> cps_bkde <- bkde2D(cbind(cps$experience, log(cps$wage)),
+    bandwidth = c(3.5, 0.5), gridsize = c(200, 200))
```

As bkde2D() does not have a formula interface (in contrast to lm() or rq()), we extract the relevant columns from the cps data set and select suitable bandwidths and grid sizes. The resulting 200×200 matrix of density estimates

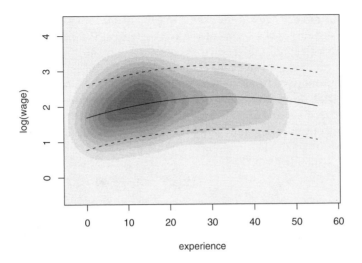

Fig. 1.4. Bivariate kernel density heatmap of log-wage by experience, with least-squares fit and prediction interval.

can be visualized in a heatmap using gray levels coding the density values. R provides `image()` (or `contour()`) to produce such displays, which can be applied to `cps_bkde` as follows.

```
R> image(cps_bkde$x1, cps_bkde$x2, cps_bkde$fhat,
+    col = rev(gray.colors(10, gamma = 1)),
+    xlab = "experience", ylab = "log(wage)")
R> box()
R> lines(fit ~ experience, data = cps2)
R> lines(lwr ~ experience, data = cps2, lty = 2)
R> lines(upr ~ experience, data = cps2, lty = 2)
```

After drawing the heatmap itself, we add the regression line for the linear model fit along with prediction intervals (see Figure 1.4). Compared with the scatterplot in Figure 1.2, this brings out more clearly the empirical relationship between `log(wage)` and experience.

This concludes our introductory R session. More details on the data sets, models, and R functions are provided in the following chapters.

1.2 Getting Started

The R system for statistical computing and graphics (R Development Core Team 2008b, http://www.R-project.org/) is an open-source software project, released under the terms of the GNU General Public License (GPL), Version 2. (Readers unfamiliar with open-source software may want to visit http://www.gnu.org/.) Its source code as well as several binary versions can be obtained, at no cost, from the Comprehensive R Archive Network (CRAN; see http://CRAN.R-project.org/mirrors.html to find its nearest mirror site). Binary versions are provided for 32-bit versions of Microsoft Windows, various flavors of Linux (including Debian, Red Hat, SUSE, and Ubuntu) and Mac OS X.

Installation

Installation of binary versions is fairly straightforward: just go to CRAN, pick the version corresponding to your operating system, and follow the instructions provided in the corresponding readme file. For Microsoft Windows, this amounts to downloading and running the setup executable (.exe file), which takes the user through a standard setup manager. For Mac OS X, separate disk image .dmg files are available for the base system as well as for a GUI developed for this platform. For various Linux flavors, there are prepackaged binaries (as .rpm or .deb files) that can be installed with the usual packaging tools on the respective platforms. Additionally, versions of R are also distributed in many of the standard Linux repositories, although these are not necessarily as quickly updated to new R versions as CRAN is.

For every system, in particular those for which a binary package does not exist, there is of course also the option to compile R from the source. On some platforms, notably Microsoft Windows, this can be a bit cumbersome because the required compilers are typically not part of a standard installation. On other platforms, such as Unix or Linux, this often just amounts to the usual configure/make/install steps. See R Development Core Team (2008d) for detailed information on installation and administration of R.

Packages

As will be discussed in greater detail below, base R is extended by means of packages, some of which are part of the default installation. Packages are stored in one or more libraries (i.e., collections of packages) on the system and can be loaded using the command library(). Typing library() without any arguments returns a list of all currently installed packages in all libraries. In the R world, some of these are referred to as "base" packages (contained in the R sources); others are "recommended" packages (included in every binary distribution). A large number of further packages, known as "contributed" packages (currently more than 1,400), are available from the CRAN servers (see http://CRAN.R-project.org/web/packages/), and some of these will be required as we proceed. Notably, the package accompanying this book, named **AER**, is needed. On a computer connected to the Internet, its installation is as simple as typing

```
R> install.packages("AER")
```

at the prompt. This installation process works on all operating systems; in addition, Windows users may install packages by using the "Install packages from CRAN" and Mac users by using the "Package installer" menu option and then choosing the packages to be installed from a list. Depending on the installation, in particular on the settings of the library paths, install.packages() by default might try to install a package in a directory where the user has no writing permission. In such a case, one needs to specify the lib argument or set the library paths appropriately (see R Development Core Team 2008d, and ?library for more information). Incidentally, installing **AER** will download several further packages on which **AER** depends. It is not uncommon for packages to depend on other packages; if this is the case, the package "knows" about it and ensures that all the functions it depends upon will become available during the installation process.

To use functions or data sets from a package, the package must be loaded. The command is, for our package **AER**,

```
R> library("AER")
```

From now on, we assume that **AER** is always loaded. It will be necessary to install and load further packages in later chapters, and it will always be indicated what they are.

In view of the rapidly increasing number of contributed packages, it has proven to be helpful to maintain a number of "CRAN task views" that provide an overview of packages for certain tasks. Current task views include econometrics, finance, social sciences, and Bayesian statistics. See http://CRAN.R-project.org/web/views/ for further details.

1.3 Working with R

There is an important difference in philosophy between R and most other econometrics packages. With many packages, an analysis will lead to a large amount of output containing information on estimation, model diagnostics, specification tests, etc. In R, an analysis is normally broken down into a series of steps. Intermediate results are stored in objects, with minimal output at each step (often none). Instead, the objects are further manipulated to obtain the information required.

In fact, the fundamental design principle underlying R (and S) is "everything is an object". Hence, not only vectors and matrices are objects that can be passed to and returned by functions, but also functions themselves, and even function calls. This enables computations on the language and can considerably facilitate programming tasks, as we will illustrate in Chapter 7.

Handling objects

To see what objects are currently defined, the function objects() (or equivalently ls()) can be used. By default, it lists all objects in the global environment (i.e., the user's workspace):

```
R> objects()
```

```
[1] "CPS1985"  "Journals" "cps"      "cps2"     "cps_bkde"
[6] "cps_lm"   "cps_rq"   "i"        "j_lm"
```

which returns a character vector of length 9 telling us that there are currently nine objects, resulting from the introductory session.

However, this cannot be the complete list of available objects, given that some objects must already exist prior to the execution of any commands, among them the function objects() that we just called. The reason is that the search list, which can be queried by

```
R> search()
```

```
 [1] ".GlobalEnv"          "package:KernSmooth"
 [3] "package:quantreg"    "package:SparseM"
 [5] "package:AER"         "package:survival"
 [7] "package:splines"     "package:strucchange"
 [9] "package:sandwich"    "package:lmtest"
[11] "package:zoo"         "package:car"
[13] "package:stats"       "package:graphics"
[15] "package:grDevices"   "package:utils"
[17] "package:datasets"    "package:methods"
[19] "Autoloads"           "package:base"
```

comprises not only the global environment ".GlobalEnv" (always at the first position) but also several attached packages, including the **base** package at its end. Calling objects("package:base") will show the names of more than a thousand objects defined in **base**, including the function objects() itself.

Objects can easily be created by assigning a value to a name using the assignment operator <-. For illustration, we create a vector x in which the number 2 is stored:

```
R> x <- 2
R> objects()
```

```
[1] "CPS1985"  "Journals" "cps"      "cps2"     "cps_bkde"
[6] "cps_lm"   "cps_rq"   "i"        "j_lm"     "x"
```

x is now available in our global environment and can be removed using the function remove() (or equivalently rm()):

```
R> remove(x)
R> objects()
```

```
[1] "CPS1985"  "Journals"  "cps"       "cps2"     "cps_bkde"
[6] "cps_lm"   "cps_rq"    "i"         "j_lm"
```

Calling functions

If the name of an object is typed at the prompt, it will be printed. For a function, say foo, this means that the corresponding R source code is printed (try, for example, objects), whereas if it is called with parentheses, as in foo(), it is a function call. If there are no arguments to the function, or all arguments have defaults (as is the case with objects()), then foo() is a valid function call. Therefore, a pair of parentheses following the object name is employed throughout this book to signal that the object discussed is a function.

Functions often have more than one argument (in fact, there is no limit to the number of arguments to R functions). A function call may use the arguments in any order, provided the name of the argument is given. If names of arguments are not given, R assumes they appear in the order of the function definition. If an argument has a default, it may be left out in a function call. For example, the function log() has two arguments, x and base: the first, x, can be a scalar (actually also a vector), the logarithm of which is to be taken; the second, base, is the base with respect to which logarithms are computed. Thus, the following four calls are all equivalent:

```
R> log(16, 2)
R> log(x = 16, 2)
R> log(16, base = 2)
R> log(base = 2, x = 16)
```

Classes and generic functions

Every object has a class that can be queried using class(). Classes include "data.frame" (a list or array with a certain structure, the preferred format in which data should be held), "lm" for linear-model objects (returned when fitting a linear regression model by ordinary least squares; see Section 1.1 above), and "matrix" (which is what the name suggests). For each class, certain methods to so-called generic functions are available; typical examples include summary() and plot(). The result of these functions depends on the class of the object: when provided with a numerical vector, summary() returns basic summaries of an empirical distribution, such as the mean and the median; for a vector of categorical data, it returns a frequency table; and in the case of a linear-model object, the result is the standard regression output. Similarly, plot() returns pairs of scatterplots when provided with a data frame and returns basic diagnostic plots for a linear-model object.

Quitting R

One exits R by using the q() function:

```
R> q()
```

R will then ask whether to save the workspace image. Answering n (no) will exit R without saving anything, whereas answering y (yes) will save all currently defined objects in a file .RData and the command history in a file .Rhistory, both in the working directory.

File management

To query the working directory, use getwd(), and to change it, setwd(). If an R session is started in a directory that has .RData and/or .Rhistory files, these will automatically be loaded. Saved workspaces from other directories can be loaded using the function load(). Analogously, R objects can be saved (in binary format) by save(). To query the files in a directory, dir() can be used.

1.4 Getting Help

R is well-documented software. Help on any function may be accessed using either ? or help(). Thus

```
R> ?options
R> help("options")
```

both open the help page for the command options(). At the bottom of a help page, there are typically practical examples of how to use that function. These can easily be executed with the example() function; e.g., example("options") or example("lm").

If the exact name of a command is not known, as will often be the case for beginners, the functions to use are help.search() and apropos(). help.search() returns help files with aliases or concepts or titles matching a "pattern" using fuzzy matching. Thus, if help on options settings is desired but the exact command name, here options(), is unknown, a search for objects containing the pattern "option" might be useful. help.search("option") will return a (long) list of commands, data frames, etc., containing this pattern, including an entry

```
options(base)        Options Settings
```

providing the desired result. It says that there exists a command options() in the **base** package that provides options settings.

Alternatively, the function apropos() lists all functions whose names include the pattern entered. As an illustration,

```
R> apropos("help")
```

```
[1] "help"        "help.search" "help.start"
```

provides a list with only three entries, including the desired command `help()`. Note that `help.search()` searches through all installed packages, whereas `apropos()` just examines the objects currently in the search list.

Vignettes

On a more advanced level, there are so-called vignettes. They are PDF files generated from integrated files containing both R code and documentation (in LaTeX format) and therefore typically contain commands that are directly executable, reproducing the analysis described. This book was written by using the tools that vignettes are based on. `vignette()` provides a list of vignettes in all attached packages. (The meaning of "attached" will be explained in Section 2.5.) As an example, `vignette("strucchange-intro", package = "strucchange")` opens the vignette accompanying the package **strucchange**. It is co-authored by the authors of this book and deals with testing, monitoring, and dating of structural changes in time series regressions. See Chapter 7 for further details on vignettes and related infrastructure.

Demos

There also exist "demos" for certain tasks. A demo is an interface to run some demonstration R scripts. Type `demo()` for a list of available topics. These include `"graphics"` and `"lm.glm"`, the latter providing illustrations on linear and generalized linear models. For beginners, running `demo("graphics")` is highly recommended.

Manuals, FAQs, and publications

R also comes with a number of manuals:

- An Introduction to R
- R Data Import/Export
- R Language Definition
- Writing R Extensions
- R Installation and Administration
- R Internals

Furthermore, there are several collections of frequently asked questions (FAQs) at `http://CRAN.R-project.org/faqs.html` that provide answers to general questions about R and also about platform-specific issues on Microsoft Windows and Mac OS X.

Moreover, there is an online newsletter named *R News*, launched in 2001. It is currently published about three times per year and features, among other things, recent developments in R (such as changes in the language or new add-on packages), a "programmer's niche", and examples analyzing data with R. See `http://CRAN.R-project.org/doc/Rnews/` for further information.

For a growing number of R packages, there exist corresponding publications in the *Journal of Statistical Software*; see `http://www.jstatsoft.org/`. This is an open-access journal that publishes articles and code snippets (as well as book and software reviews) on the subject of statistical software and algorithms. A special volume on *Econometrics in R* is currently in preparation.

Finally, there is a rapidly growing list of books on R, or on statistics using R, at all levels, the most comprehensive perhaps being Venables and Ripley (2002). In addition, we refer the interested reader to Dalgaard (2002) for introductory statistics, to Murrell (2005) for R graphics, and to Faraway (2005) for linear regression.

1.5 The Development Model

One of R's strengths and a key feature in its success is that it is highly extensible through packages that can provide extensions to everything available in the base system. This includes not only R code but also code in compiled languages (such as C, C++, or FORTRAN), data sets, demo files, test suites, vignettes, or further documentation. Therefore, every R user can easily become an R developer by submitting his or her packages to CRAN to share them with the R community. Hence packages can actively influence the direction in which (parts of) R will go in the future.

Unlike the CRAN packages, the base R system is maintained and developed only by the R core team, which releases major version updates (i.e., versions x.y.0) biannually (currently around April 1 and October 1). However, as R is an open-source system, all users are given read access to the master SVN repository—SVN stands for Subversion and is a version control system; see `http://subversion.tigris.org/`—and can thus check out the full source code of the development version of R.

In addition, there are several means of communication within the R user and developer community and between the users and the core development team. The two most important are various R mailing lists and, as described above, CRAN packages. The R project hosts several mailing lists, including R-help and R-devel. The former can be used to ask for help on using R, the latter for discussing issues related to the development of R or R packages. Furthermore, bugs can be reported and feature requests made. The posting guide at `http://www.R-project.org/posting-guide.html` discusses some good strategies for doing this effectively. In addition to these general mailing lists, there are lists for special interest groups (SIGs), among them at least one list that might be of interest to the reader: it is devoted to finance and (financial) econometrics and is called R-SIG-Finance.

1.6 A Brief History of R

As noted above, the R system for statistical computing and graphics (R Development Core Team 2008b, http://www.R-project.org/) is an open-source software project. The story begins at Bell Laboratories (of AT&T and now Alcatel-Lucent in New Jersey), with the S language, a system for data analysis developed by John Chambers, Rick Becker, and collaborators since the late 1970s. Landmarks of the development of S are a series of books, referred to by color in the S community, beginning with the "brown book" (Becker and Chambers 1984), which features "Old S". The basic reference for "New S", or S version 2, is Becker, Chambers, and Wilks (1988), the "blue book". For S version 3 (first-generation object-oriented programming and statistical modeling), it is Chambers and Hastie (1992), the "white book". The "green book" (Chambers 1998) documents S version 4. Based on the various S versions, Insightful Corporation (formerly MathSoft and still earlier Statistical Sciences) has provided a commercially enhanced and supported release of S, named S-PLUS, since 1987. At its core, this includes the original S implementation, which was first exclusively licensed and finally purchased in 2004. On March 23, 1999, the Association for Computing Machinery (ACM) named John Chambers as the recipient of its 1998 Software System Award for developing the S system, noting that his work "will forever alter the way people analyze, visualize, and manipulate data".

R itself was initially developed by Robert Gentleman and Ross Ihaka at the University of Auckland, New Zealand. An early version is described in an article by its inventors (Ihaka and Gentleman 1996). They designed the language to combine the strengths of two existing languages, S and Scheme (Steel and Sussman 1975). In the words of Ihaka and Gentleman (1996), "[t]he resulting language is very similar in appearance to S, but the underlying implementation and semantics are derived from Scheme". The result was baptized R "in part to acknowledge the influence of S and in part to celebrate [their] own efforts".

The R source code was first released under the GNU General Public License (GPL) in 1995. Since mid-1997, there has been the R Development Core Team, currently comprising 19 members, and their names are available upon typing contributors() in an R session. In 1998, the Comprehensive R Archive Network (CRAN; http://CRAN.R-project.org/) was established, which is a family of mirror sites around the world that store identical, up-to-date versions of code and documentation for R. The first official release, R version 1.0.0, dates to 2000-02-29. It implements the S3 standard as documented by Chambers and Hastie (1992). R version 2.0.0 was released in 2004, and the current version of the language, R 2.7.0, may be viewed as implementing S4 (Chambers 1998) plus numerous concepts that go beyond the various S standards.

The first publication on R in the econometrics literature appears to have been by Cribari-Neto and Zarkos (1999), a software review in the *Journal of Applied Econometrics* entitled "R: Yet Another Econometric Programming Environment". It describes R version 0.63.1, still a beta version. Three years later, in a further software review in the same journal, Racine and Hyndman (2002) focused on using R for teaching econometrics utilizing R 1.3.1. To the best of our knowledge, this book is the first general introduction to econometric computing with R.

2

Basics

R can be used at various levels. Of course, standard arithmetic is available, and hence it can be used as a (rather sophisticated) calculator. It is also provided with a graphical system that writes on a large variety of devices. Furthermore, R is a full-featured programming language that can be employed to tackle all typical tasks for which other programming languages are also used. It connects to other languages, programs, and data bases, and also to the operating system; users can control all these from within R.

In this chapter, we illustrate a few of the typical uses of R. Often solutions are not unique, but in the following we avoid sophisticated shortcuts. However, we encourage all readers to explore alternative solutions by reusing what they have learned in other contexts.

2.1 R as a Calculator

The standard arithmetic operators +, -, *, /, and ^ are available, where x^y yields x^y. Hence

```
R> 1 + 1

[1] 2

R> 2^3

[1] 8
```

In the output, [1] indicates the position of the first element of the vector returned by R. This is not surprising here, where all vectors are of length 1, but it will be useful later.

The common mathematical functions, such as log(), exp(), sin(), asin(), cos(), acos(), tan(), atan(), sign(), sqrt(), abs(), min(), and max(), are also available. Specifically, log(x, base = a) returns the logarithm of x to base a, where a defaults to exp(1). Thus

C. Kleiber, A. Zeileis, *Applied Econometrics with R*,
DOI: 10.1007/978-0-387-77318-6_2, © Springer Science+Business Media, LLC 2008

```
R> log(exp(sin(pi/4)^2) * exp(cos(pi/4)^2))
```

```
[1] 1
```

which also shows that `pi` is a built-in constant. There are further convenience functions, such as `log10()` and `log2()`, but here we shall mainly use `log()`. A full list of all options and related functions is available upon typing `?log`, `?sin`, etc. Additional functions useful in statistics and econometrics are `gamma()`, `beta()`, and their logarithms and derivatives. See `?gamma` for further information.

Vector arithmetic

In R, the basic unit is a vector, and hence all these functions operate directly on vectors. A vector is generated using the function `c()`, where `c` stands for "combine" or "concatenate". Thus

```
R> x <- c(1.8, 3.14, 4, 88.169, 13)
```

generates an object x, a vector, containing the entries 1.8, 3.14, 4, 88.169, 13. The length of a vector is available using `length()`; thus

```
R> length(x)
```

```
[1] 5
```

Note that names are case-sensitive; hence x and X are distinct.

The preceding statement uses the assignment operator `<-`, which should be read as a single symbol (although it requires two keystrokes), an arrow pointing to the variable to which the value is assigned. Alternatively, `=` may be used at the user level, but since `<-` is preferred for programming, it is used throughout this book. There is no immediately visible result, but from now on x has as its value the vector defined above, and hence it can be used in subsequent computations:

```
R> 2 * x + 3
```

```
[1]    6.60    9.28   11.00 179.34   29.00
```

```
R> 5:1 * x + 1:5
```

```
[1]   10.00   14.56   15.00 180.34   18.00
```

This requires an explanation. In the first statement, the scalars (i.e., vectors of length 1) 2 and 3 are recycled to the length of x so that each element of x is multiplied by 2 before 3 is added. In the second statement, x is multiplied element-wise by the vector 1:5 (the sequence from 1 to 5; see below) and then the vector 5:1 is added element-wise.

Mathematical functions can be applied as well; thus

```
R> log(x)
```

```
[1] 0.5878 1.1442 1.3863 4.4793 2.5649
```

returns a vector containing the logarithms of the original entries of x.

Subsetting vectors

It is often necessary to access subsets of vectors. This requires the operator
[, which can be used in several ways to extract elements of a vector. For
example, one can either specify which elements to include or which elements
to exclude: a vector of positive indices, such as

```
R> x[c(1, 4)]
```

```
[1]    1.80 88.17
```

specifies the elements to be extracted. Alternatively, a vector of negative indices, as in

```
R> x[-c(2, 3, 5)]
```

```
[1]    1.80 88.17
```

selects all elements but those indicated, yielding the same result. In fact,
further methods are available for subsetting with [, which are explained below.

Patterned vectors

In statistics and econometrics, there are many instances where vectors with
special patterns are needed. R provides a number of functions for creating
such vectors, including

```
R> ones <- rep(1, 10)
R> even <- seq(from = 2, to = 20, by = 2)
R> trend <- 1981:2005
```

Here, ones is a vector of ones of length 10, even is a vector containing the
even numbers from 2 to 20, and trend is a vector containing the integers from
1981 to 2005.

Since the basic element is a vector, it is also possible to concatenate vectors.
Thus

```
R> c(ones, even)
```

```
[1]    1  1  1  1  1  1  1  1  1  1  2  4  6  8 10 12 14 16 18 20
```

creates a vector of length 20 consisting of the previously defined vectors ones
and even laid end to end.

2.2 Matrix Operations

A 2×3 matrix containing the elements 1:6, by column, is generated via

```
R> A <- matrix(1:6, nrow = 2)
```

Alternatively, ncol could have been used, with matrix(1:6, ncol = 3) yielding the same result.

Basic matrix algebra

The transpose $A^\top$ of A is

```
R> t(A)
```

```
     [,1] [,2]
[1,]   1    2
[2,]   3    4
[3,]   5    6
```

The dimensions of a matrix may be accessed using dim(), nrow(), and ncol(); hence

```
R> dim(A)
```

```
[1] 2 3
```

```
R> nrow(A)
```

```
[1] 2
```

```
R> ncol(A)
```

```
[1] 3
```

Single elements of a matrix, row or column vectors, or indeed entire submatrices may be extracted by specifying the rows and columns of the matrix from which they are selected. This uses a simple extension of the rules for subsetting vectors. (In fact, internally, matrices are vectors with an additional dimension attribute enabling row/column-type indexing.) Element a_{ij} of a matrix A is extracted using A[i,j]. Entire rows or columns can be extracted via A[i,] and A[,j], respectively, which return the corresponding row or column *vectors*. This means that the dimension attribute is dropped (by default); hence subsetting will return a vector instead of a matrix if the resulting matrix has only one column or row. Occasionally, it is necessary to extract rows, columns, or even single elements of a matrix as a matrix. Dropping of the dimension attribute can be switched off using A[i, j, drop = FALSE]. As an example,

```
R> A1 <- A[1:2, c(1, 3)]
```

selects a square matrix containing the first and third elements from each row (note that A has only two rows in our example). Alternatively, and more compactly, A1 could have been generated using A[, -2]. If no row number is specified, all rows will be taken; the -2 specifies that all columns but the second are required.

A1 is a square matrix, and if it is nonsingular it has an inverse. One way to check for singularity is to compute the determinant using the R function det(). Here, det(A1) equals -4; hence A1 is nonsingular. Alternatively, its eigenvalues (and eigenvectors) are available using eigen(). Here, eigen(A1) yields the eigenvalues 7.531 and -0.531, again showing that A1 is nonsingular.

The inverse of a matrix, if it cannot be avoided, is computed using solve():

```
R> solve(A1)
```

```
     [,1]  [,2]
[1,] -1.5  1.25
[2,]  0.5 -0.25
```

We can check that this is indeed the inverse of A1 by multiplying A1 with its inverse. This requires the operator for matrix multiplication, %*%:

```
R> A1 %*% solve(A1)
```

```
     [,1] [,2]
[1,]   1    0
[2,]   0    1
```

Similarly, conformable matrices are added and subtracted using the arithmetical operators + and -. It is worth noting that for non-conformable matrices recycling proceeds along columns. Incidentally, the operator * also works for matrices; it returns the element-wise or Hadamard product of conformable matrices. Further types of matrix products that are often required in econometrics are the Kronecker product, available via kronecker(), and the cross product $A^\top B$ of two matrices, for which a computationally efficient algorithm is implemented in crossprod().

In addition to the spectral decomposition computed by eigen() as mentioned above, R provides other frequently used matrix decompositions, including the singular-value decomposition svd(), the QR decomposition qr(), and the Cholesky decomposition chol().

Patterned matrices

In econometrics, there are many instances where matrices with special patterns occur. R provides functions for generating matrices with such patterns. For example, a diagonal matrix with ones on the diagonal may be created using

```
R> diag(4)
```

```
     [,1] [,2] [,3] [,4]
[1,]   1    0    0    0
[2,]   0    1    0    0
[3,]   0    0    1    0
[4,]   0    0    0    1
```

which yields the 4×4 identity matrix. Equivalently, it can be obtained by diag(1, 4, 4), where the 1 is recycled to the required length 4. Of course, more general diagonal matrices are also easily obtained: diag(rep(c(1,2), c(10, 10))) yields a diagonal matrix of size 20×20 whose first 10 diagonal elements are equal to 1, while the remaining ones are equal to 2. (Readers with a basic knowledge of linear regression will note that an application could be a pattern of heteroskedasticity.)

Apart from setting up diagonal matrices, the function diag() can also be used for extracting the diagonal from an existing matrix; e.g., diag(A1). Additionally, upper.tri() and lower.tri() can be used to query the positions of upper or lower triangular elements of a matrix, respectively.

Combining matrices

It is also possible to form new matrices from existing ones. This uses the functions rbind() and cbind(), which are similar to the function c() for concatenating vectors; as their names suggest, they combine matrices by rows or columns. For example, to add a column of ones to our matrix A1,

R> cbind(1, A1)

```
     [,1] [,2] [,3]
[1,]   1    1    5
[2,]   1    2    6
```

can be employed, while

R> rbind(A1, diag(4, 2))

```
     [,1] [,2]
[1,]   1    5
[2,]   2    6
[3,]   4    0
[4,]   0    4
```

combines A1 and diag(4, 2) by rows.

2.3 R as a Programming Language

R is a full-featured, interpreted, object-oriented programming language. Hence, it can be used for all the tasks other programming languages are also used for, not only for data analysis. What makes it particularly useful for statistics and econometrics is that it was designed for "programming with data" (Chambers 1998). This has several implications for the data types employed and the object-orientation paradigm used (see Section 2.6 for more on object orientation).

An in-depth treatment of programming in S/R is given in Venables and Ripley (2000). If you read German, Ligges (2007) is an excellent introduction to programming with R. On a more advanced level, R Development Core Team (2008f,g) provides guidance about the language definition and how extensions to the R system can be written. The latter documents can be downloaded from CRAN and also ship with every distribution of R.

The mode of a vector

Probably the simplest data structure in R is a vector. All elements of a vector must be of the same type; technically, they must be of the same "mode". The mode of a vector x can be queried using mode(x). Here, we need vectors of modes "numeric", "logical", and "character" (but there are others).

We have already seen that

```
R> x <- c(1.8, 3.14, 4, 88.169, 13)
```

creates a numerical vector, and one typical application of vectors is to store the values of some numerical variable in a data set.

Logical and character vectors

Logical vectors may contain the logical constants TRUE and FALSE. In a fresh session, the aliases T and F are also available for compatibility with S (which uses these as the logical constants). However, unlike TRUE and FALSE, the values of T and F can be changed (e.g., by using the former for signifying the sample size in a time series context or using the latter as the variable for an F statistic), and hence it is recommended not to rely on them but to always use TRUE and FALSE. Like numerical vectors, logical vectors can be created from scratch. They may also arise as the result of a logical comparison:

```
R> x > 3.5
```

```
[1] FALSE FALSE  TRUE  TRUE  TRUE
```

Further logical operations are explained below.

Character vectors can be employed to store strings. Especially in the early chapters of this book, we will mainly use them to assign labels or names to certain objects such as vectors and matrices. For example, we can assign names to the elements of x via

```
R> names(x) <- c("a", "b", "c", "d", "e")
R> x
```

```
   a     b     c     d     e
1.80  3.14  4.00 88.17 13.00
```

Alternatively, we could have used `names(x) <- letters[1:5]` since `letters` and `LETTERS` are built-in vectors containing the 26 lower- and uppercase letters of the Latin alphabet, respectively. Although we do not make much use of them in this book, we note here that the character-manipulation facilities of R go far beyond these simple examples, allowing, among other things, computations on text documents or command strings.

More on subsetting

Having introduced vectors of modes numeric, character, and logical, it is useful to revisit subsetting of vectors. By now, we have seen how to extract parts of a vector using numerical indices, but in fact this is also possible using characters (if there is a `names` attribute) or logicals (in which case the elements corresponding to `TRUE` are selected). Hence, the following commands yield the same result:

```
R> x[3:5]
```

```
   c     d     e
4.00 88.17 13.00
```

```
R> x[c("c", "d", "e")]
```

```
   c     d     e
4.00 88.17 13.00
```

```
R> x[x > 3.5]
```

```
   c     d     e
4.00 88.17 13.00
```

Subsetting of matrices (and also of data frames or multidimensional arrays) works similarly.

Lists

So far, we have only used plain vectors. We now proceed to introduce some related data structures that are similar but contain more information.

In R, lists are *generic vectors* where each element can be virtually any type of object; e.g., a vector (of arbitrary mode), a matrix, a full data frame, a function, or again a list. Note that the latter also allows us to create recursive data structures. Due to this flexibility, lists are the basis for most complex objects in R; e.g., for data frames or fitted regression models (to name two examples that will be described later).

As a simple example, we create, using the function `list()`, a list containing a sample from a standard normal distribution (generated with `rnorm()`; see below) plus some markup in the form of a character string and a list containing the population parameters.

```
R> mylist <- list(sample = rnorm(5),
+     family = "normal distribution",
+     parameters = list(mean = 0, sd = 1))
R> mylist

$sample
[1]  0.3771 -0.9346  2.4302  1.3195  0.4503

$family
[1] "normal distribution"

$parameters
$parameters$mean
[1] 0

$parameters$sd
[1] 1
```

To select certain elements from a list, the extraction operators $ or [[can be used. The latter is similar to [, the main difference being that it can only select a single element. Hence, the following statements are equivalent:

```
R> mylist[[1]]

[1]  0.3771 -0.9346  2.4302  1.3195  0.4503

R> mylist[["sample"]]

[1]  0.3771 -0.9346  2.4302  1.3195  0.4503

R> mylist$sample

[1]  0.3771 -0.9346  2.4302  1.3195  0.4503
```

The third element of mylist again being a list, the extractor functions can also be combined as in

```
R> mylist[[3]]$sd

[1] 1
```

Logical comparisons

R has a set of functions implementing standard logical comparisons as well as a few further functions that are convenient when working with logical values. The logical operators are <, <=, >, >=, == (for exact equality) and != (for inequality). Also, if expr1 and expr2 are logical expressions, then expr1 & expr2 is their intersection (logical "and"), expr1 | expr2 is their union (logical "or"), and !expr1 is the negation of expr1. Thus

```
R> x <- c(1.8, 3.14, 4, 88.169, 13)
R> x > 3 & x <= 4
```

```
[1] FALSE  TRUE  TRUE FALSE FALSE
```

To assess for which elements of a vector a certain expression is TRUE, the function which() can be used:

```
R> which(x > 3 & x <= 4)
```

```
[1] 2 3
```

The specialized functions which.min() and which.max() are available for computing the position of the minimum and the maximum. In addition to & and |, the functions all() and any() check whether all or at least some entries of a vector are TRUE:

```
R> all(x > 3)
```

```
[1] FALSE
```

```
R> any(x > 3)
```

```
[1] TRUE
```

Some caution is needed when assessing exact equality. When applied to numerical input, == does not allow for finite representation of fractions or for rounding error; hence situations like

```
R> (1.5 - 0.5) == 1
```

```
[1] TRUE
```

```
R> (1.9 - 0.9) == 1
```

```
[1] FALSE
```

can occur due to floating-point arithmetic (Goldberg 1991). For these purposes, all.equal() is preferred:

```
R> all.equal(1.9 - 0.9, 1)
```

```
[1] TRUE
```

Furthermore, the function identical() checks whether two R objects are exactly identical.

Due to coercion, it is also possible to compute directly on logical vectors using ordinary arithmetic. When coerced to numeric, FALSE becomes 0 and TRUE becomes 1, as in

```
R> 7 + TRUE
```

```
[1] 8
```

Coercion

To convert an object from one type or class to a different one, R provides a number of coercion functions, conventionally named as.*foo* (), where *foo* is the desired type or class; e.g., numeric, character, matrix, and data.frame (a concept introduced in Section 2.5), among many others. They are typically accompanied by an is.*foo* () function that checks whether an object is of type or class *foo*. Thus

```
R> is.numeric(x)
```

```
[1] TRUE
```

```
R> is.character(x)
```

```
[1] FALSE
```

```
R> as.character(x)
```

```
[1] "1.8"    "3.14"   "4"      "88.169" "13"
```

In certain situations, coercion is also forced automatically by R; e.g., when the user tries to put elements of different modes into a single vector (which can only contain elements of the same mode). Thus

```
R> c(1, "a")
```

```
[1] "1" "a"
```

Random number generation

For programming environments in statistics and econometrics, it is vital to have good random number generators (RNGs) available, in particular to allow the users to carry out Monte Carlo studies. The R RNG supports several algorithms; see ?RNG for further details. Here, we outline a few important commands.

The RNG relies on a "random seed", which is the basis for the generation of pseudo-random numbers. By setting the seed to a specific value using the function set.seed(), simulations can be made exactly reproducible. For example, using the function rnorm() for generating normal random numbers,

```
R> set.seed(123)
R> rnorm(2)
```

```
[1] -0.5605 -0.2302
```

```
R> rnorm(2)
```

```
[1] 1.55871 0.07051
```

```
R> set.seed(123)
R> rnorm(2)
```

```
[1] -0.5605 -0.2302
```

Another basic function for drawing random samples, with or without replacement from a finite set of values, is sample(). The default is to draw, without
replacement, a vector of the same size as its input argument; i.e., to compute
a permutation of the input as in

```
R> sample(1:5)
```

```
[1] 5 1 2 3 4
```

```
R> sample(c("male", "female"), size = 5, replace = TRUE,
+    prob = c(0.2, 0.8))
```

```
[1] "female" "male"   "female" "female" "female"
```

The second command draws a sample of size 5, with replacement, from the
values "male" and "female", which are drawn with probabilities 0.2 and 0.8,
respectively.

Above, we have already used the function rnorm() for drawing from a
normal distribution. It belongs to a broader family of functions that are all of
the form r*dist*(), where *dist* can be, for example, norm, unif, binom, pois, t,
f, chisq, corresponding to the obvious families of distributions. All of these
functions take the sample size n as their first argument along with further
arguments controlling parameters of the respective distribution. For example, rnorm() takes mean and sd as further arguments, with 0 and 1 being
the corresponding defaults. However, these are not the only functions available for statistical distributions. Typically there also exist d*dist*(), p*dist*(),
and q*dist*(), which implement the density, cumulative probability distribution
function, and quantile function (inverse distribution function), respectively.

Flow control

Like most programming languages, R provides standard control structures
such as if/else statements, for loops, and while loops. All of these have in
common that an expression expr is evaluated, either conditional upon a certain condition cond (for if and while) or for a sequence of values (for for).
The expression expr itself can be either a simple expression or a compound expression; i.e., typically a set of simple expressions enclosed in braces { ... }.
Below we present a few brief examples illustrating its use; see ?Control for
further information.

An if/else statement is of the form

```
if(cond) {
  expr1
} else {
  expr2
}
```

where `expr1` is evaluated if `cond` is `TRUE` and `expr2` otherwise. The `else` branch may be omitted if empty. A simple (if not very meaningful) example is

```
R> x <- c(1.8, 3.14, 4, 88.169, 13)
R> if(rnorm(1) > 0) sum(x) else mean(x)
```

```
[1] 22.02
```

where conditional on the value of a standard normal random number either the sum or the mean of the vector `x` is computed. Note that the condition `cond` can only be of length 1. However, there is also a function `ifelse()` offering a vectorized version; e.g.,

```
R> ifelse(x > 4, sqrt(x), x^2)
```

```
[1]  3.240  9.860 16.000  9.390  3.606
```

This computes the square root for those values in `x` that are greater than 4 and the square for the remaining ones.

A `for` loop looks similar, but the main argument to `for()` is of type `variable in sequence`. To illustrate its use, we recursively compute first differences in the vector `x`.

```
R> for(i in 2:5) {
+     x[i] <- x[i] - x[i-1]
+ }
R> x[-1]
```

```
[1]   1.34   2.66  85.51 -72.51
```

Finally, a `while` loop looks quite similar. The argument to `while()` is a condition that may change in every run of the loop so that it finally can become `FALSE`, as in

```
R> while(sum(x) < 100) {
+     x <- 2 * x
+ }
R> x
```

```
[1]   14.40   10.72   21.28  684.07 -580.07
```

Writing functions

One of the features of S and R is that users naturally become developers. Creating variables or objects and applying functions to them interactively (either to modify them or to create other objects of interest) is part of every R session. In doing so, typical sequences of commands often emerge that are carried out for different sets of input values. Instead of repeating the same steps "by hand", they can also be easily wrapped into a function. A simple example is

```
R> cmeans <- function(X) {
+    rval <- rep(0, ncol(X))
+    for(j in 1:ncol(X)) {
+      mysum <- 0
+      for(i in 1:nrow(X)) mysum <- mysum + X[i,j]
+      rval[j] <- mysum/nrow(X)
+    }
+    return(rval)
+  }
```

This creates a (deliberately awkward!) function cmeans(), which takes a matrix argument X and uses a double for loop to compute first the sum and then the mean in each column. The result is stored in a vector rval (our return value), which is returned after both loops are completed. This function can then be easily applied to new data, as in

```
R> X <- matrix(1:20, ncol = 2)
R> cmeans(X)
```

```
[1]  5.5 15.5
```

and (not surprisingly) yields the same result as the built-in function colMeans():

```
R> colMeans(X)
```

```
[1]  5.5 15.5
```

The function cmeans() takes only a single argument X that has no default value. If the author of a function wants to set a default, this can be easily achieved by defining a function with a list of name = expr pairs, where name is the argument of the variable and expr is an expression with the default value. If the latter is omitted, no default value is set.

In interpreted matrix-based languages such as R, loops are typically less efficient than the corresponding vectorized computations offered by the system. Therefore, avoiding loops by replacing them with vectorized operations can save computation time, especially when the number of iterations in the loop can become large. To illustrate, let us generate $2 \cdot 10^6$ random numbers from the standard normal distribution and compare the built-in function colMeans() with our awkward function cmeans(). We employ the function system.time(), which is useful for profiling code:

```
R> X <- matrix(rnorm(2*10^6), ncol = 2)
R> system.time(colMeans(X))
```

```
   user  system elapsed
  0.004   0.000   0.005
```

```
R> system.time(cmeans(X))
```

```
  user  system elapsed
 5.572   0.004   5.617
```

Clearly, the performance of cmeans() is embarrassing, and using colMeans()
is preferred.

Vectorized calculations

As noted above, loops can be avoided using vectorized arithmetic. In the
case of cmeans(), our function calculating column-wise means of a matrix,
it would be helpful to directly compute means column by column using the
built-in function mean(). This is indeed the preferred solution. Using the tools
available to us thus far, we could proceed as follows:

```
R> cmeans2 <- function(X) {
+     rval <- rep(0, ncol(X))
+     for(j in 1:ncol(X)) rval[j] <- mean(X[,j])
+     return(rval)
+ }
```

This eliminates one of the for loops and only cycles over the columns. The
result is identical to the previous solutions, but the performance is clearly
better than that of cmeans():

```
R> system.time(cmeans2(X))
```

```
  user  system elapsed
 0.072   0.008   0.080
```

However, the code of cmeans2() still looks a bit cumbersome with the remaining for loop—it can be written much more compactly using the function
apply(). This applies functions over the margins of an array and takes three
arguments: the array, the index of the margin, and the function to be evaluated. In our case, the function call is

```
R> apply(X, 2, mean)
```

because we require means (using mean()) over the columns (i.e., the second
dimension) of X. The performance of apply() can sometimes be better than
a for loop; however, in many cases both approaches perform rather similarly:

```
R> system.time(apply(X, 2, mean))
```

```
  user  system elapsed
 0.084   0.028   0.114
```

To summarize, this means that (1) element-wise computations should be
avoided if vectorized computations are available, (2) optimized solutions (if
available) typically perform better than the generic for or apply() solution,
and (3) loops can be written more compactly using the apply() function. In

fact, this is so common in R that several variants of `apply()` are available, namely `lapply()`, `tapply()`, and `sapply()`. The first returns a list, the second a table, and the third tries to simplify the result to a vector or matrix where possible. See the corresponding manual pages for more detailed information and examples.

Reserved words

Like most programming languages, R has a number of reserved words that provide the basic grammatical constructs of the language. Some of these have already been introduced above, and some more follow below. An almost complete list of reserved words in R is: `if`, `else`, `for`, `in`, `while`, `repeat`, `break`, `next`, `function`, `TRUE`, `FALSE`, `NA`, `NULL`, `Inf`, `NaN`, ...). See `?Reserved` for a complete list. If it is attempted to use any of these as names, this results in an error.

2.4 Formulas

Formulas are constructs used in various statistical programs for specifying models. In R, formula objects can be used for storing symbolic descriptions of relationships among variables, such as the ~ operator in the formation of a formula:

```
R> f <- y ~ x
R> class(f)
```

```
[1] "formula"
```

So far, this is only a description without any concrete meaning. The result entirely depends on the function evaluating this formula. In R, the expression above commonly means "y is explained by x". Such formula interfaces are convenient for specifying, among other things, plots or regression relationships. For example, with

```
R> x <- seq(from = 0, to = 10, by = 0.5)
R> y <- 2 + 3 * x + rnorm(21)
```

the code

```
R> plot(y ~ x)
R> lm(y ~ x)
```

```
Call:
lm(formula = y ~ x)

Coefficients:
(Intercept)            x
       2.00         3.01
```

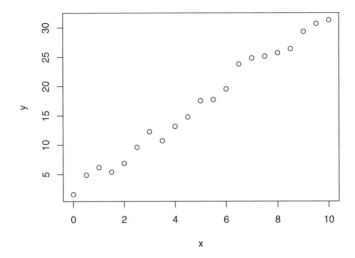

Fig. 2.1. Simple scatterplot of y vs. x.

first generates a scatterplot of y against x (see Figure 2.1) and then fits the corresponding simple linear regression model with slope 3.01 and intercept 2.00.

For specifying regression models, the formula language is much richer than outlined above and is based on a symbolic notation suggested by Wilkinson and Rogers (1973) in the statistical literature. For example, when using lm(), log(y) ~ x1 + x2 specifies a linear regression of log(y) on two regressors x1 and x2 and an implicitly defined constant. More details on the formula specifications of linear regression models will be given in Chapter 3.

2.5 Data Management in R

In R, a data frame corresponds to what other statistical packages call a data matrix or a data set. Typically, it is an array consisting of a list of vectors and/or factors of identical length, thus yielding a rectangular format where columns correspond to variables and rows to observations.

Creation from scratch

Let us generate a simple artificial data set, with three variables named "one", "two", "three", by using

```
R> mydata <- data.frame(one = 1:10, two = 11:20, three = 21:30)
```

Alternatively, `mydata` can be created using

```
R> mydata <- as.data.frame(matrix(1:30, ncol = 3))
R> names(mydata) <- c("one", "two", "three")
```

which first creates a matrix of size 10×3 that is subsequently coerced to a data frame and whose variable names are finally changed to "one", "two", "three". Note that the same syntax can be used both for querying and modifying the names in a data frame. Furthermore, it is worth reiterating that although a data frame can be coerced from a matrix as above, it is internally represented as a list.

Subset selection

It is possible to access a subset of variables (i.e., columns) via [or $, where the latter can only extract a single variable. Hence, the second variable `two` can be selected via

```
R> mydata$two
```

```
 [1] 11 12 13 14 15 16 17 18 19 20
```

```
R> mydata[, "two"]
```

```
 [1] 11 12 13 14 15 16 17 18 19 20
```

```
R> mydata[, 2]
```

```
 [1] 11 12 13 14 15 16 17 18 19 20
```

In all cases, the object returned is a simple vector; i.e., the data frame attributes are dropped (by default).

To simplify access to variables in a certain data set, it can be `attach()`ed. Technically, this means that the attached data set is added to the `search()` path and thus variables contained in this data set can be found when their name is used in a command. Compare the following:

```
R> mean(two)
```

```
Error in mean(two) : Object "two" not found
```

```
R> attach(mydata)
R> mean(two)
```

```
[1] 15.5
```

```
R> detach(mydata)
```

Data frames should be attached with care; in particular, one should pay attention not to attach several data frames with the same column names or to have a variable with identical name in the global environment, as this is likely to generate confusion. To avoid attaching and detaching a data set for a single command only, the function with() can be handy, as in

```
R> with(mydata, mean(two))
```

```
[1] 15.5
```

It is often necessary to work with subsets of a data frame; i.e., to use only selected observations (= rows) and/or variables (= columns). This can again be done via [or, more conveniently, using the subset() command, whose main arguments are a data frame from which the subset is to be taken and a logical statement defining the elements to be selected. For example,

```
R> mydata.sub <- subset(mydata, two <= 16, select = -two)
```

takes all observations whose value of the second variable two does not exceed 16 (we know there are six observations with this property) and, in addition, all variables apart from two are selected.

Import and export

To export data frames in plain text format, the function write.table() can be employed:

```
R> write.table(mydata, file = "mydata.txt", col.names = TRUE)
```

It creates a text file mydata.txt in the current working directory. If this data set is to be used again, in another session, it may be imported using

```
R> newdata <- read.table("mydata.txt", header = TRUE)
```

The function read.table() returns a "data.frame" object, which is then assigned to a new object newdata. By setting col.names = TRUE, the column names are written in the first line of mydata.txt and hence we set header = TRUE when reading the file again. The function write.table() is quite flexible and allows specification of the separation symbol and the decimal separator, among other properties of the file to be written, so that various text-based formats, including tab- or comma-separated values, can be produced. Since the latter is a popular format for exchanging data (as it can be read and written by many spreadsheet programs, including Microsoft Excel), the convenience interfaces read.csv() and write.csv() are available. Similarly, read.csv2() and write.csv2() provide export and import of semicolon-separated values, a format that is typically used on systems employing the comma (and not the period) as the decimal separator. In addition, there exists a more elementary function, named scan(), for data not conforming to the matrix-like layout required by read.table(). We refer to the respective manual pages and the "R

Data Import/Export" manual (R Development Core Team 2008c) for further details.

It is also possible to save the data in the R internal binary format, by convention with extension `.RData` or `.rda`. The command

```
R> save(mydata, file = "mydata.rda")
```

saves the data in R binary format. Binary files may be loaded using

```
R> load("mydata.rda")
```

In contrast to `read.table()`, this does not return a single object; instead it makes all objects stored in `mydata.rda` directly available within the current environment. The advantage of using `.rda` files is that several R objects, in fact several arbitrary R objects, can be stored, including functions or fitted models, without loss of information.

All of the data sets in the package **AER** are supplied in this binary format (go to the folder `~/AER/data` in your R library to check). Since they are part of a package, they are made accessible more easily using `data()` (which in this case sets up the appropriate call for `load()`). Thus

```
R> data("Journals", package = "AER")
```

loads the `Journals` data frame from the **AER** package (stored in the file `~/AER/data/Journals.rda`), the data set used in Example 1 of our introductory R session. If the `package` argument is omitted, all packages currently in the search path are checked whether they provide a data set `Journals`.

Reading and writing foreign binary formats

R can also read and write a number of proprietary binary formats, notably S-PLUS, SPSS, SAS, Stata, Minitab, Systat, and dBase files, using the functions provided in the package **foreign** (part of a standard R installation). Most of the commands are designed to be similar to `read.table()` and `write.table()`. For example, for Stata files, both `read.dta()` and `write.dta()` are available and can be used to create a Stata file containing `mydata`

```
R> library("foreign")
R> write.dta(mydata, file = "mydata.dta")
```

and read it into R again via

```
R> mydata <- read.dta("mydata.dta")
```

See the documentation for the package **foreign** for further information.

Interaction with the file system and string manipulations

In the preceding paragraphs, some interaction with the file system was necessary to read and write data files. R possesses a rich functionality for interacting with external files and communicating with the operating system. This is far beyond the scope of this book, but we would like to provide the interested reader with a few pointers that may serve as a basis for further reading.

Files available in a directory or folder can be queried via dir() and also copied (using file.copy()) or deleted (using file.remove()) independent of the operating system. For example, the Stata file created above can be deleted again from within R via

```
R> file.remove("mydata.dta")
```

Other (potentially system-dependent) commands can be sent as strings to the operating system using system(). See the respective manual pages for more information and worked-out examples.

Above, we discussed how data objects (especially data frames) can be written to files in various formats. Beyond that, one often wants to save commands or their output to text files. One possibility to achieve this is to use sink(), which can direct output to a file() connection to which the strings could be written with cat(). In some situations, writeLines() is more convenient for this. Furthermore, dump() can create text representations of R objects and write them to a file() connection.

Sometimes, one needs to manipulate the strings before creating output. R also provides rich and flexible functionality for this. Typical tasks include splitting strings (strsplit()) and/or pasting them together (paste()). For pattern matching and replacing, grep() and gsub() are available, which also support regular expressions. For combining text and variable values, sprintf() is helpful.

Factors

Factors are an extension of vectors designed for storing categorical information. Typical econometric examples of categorical variables include gender, union membership, or ethnicity. In many software packages, these are created using a numerical encoding (e.g., 0 for males and 1 for females); sometimes, especially in regression settings, a single categorical variable is stored in several such dummy variables if there are more than two categories.

In R, categorical variables should be specified as factors. As an example, we first create a dummy-coded vector with a certain pattern and subsequently transform it into a factor using factor():

```
R> g <- rep(0:1, c(2, 4))
R> g <- factor(g, levels = 0:1, labels = c("male", "female"))
R> g

[1] male    male    female female female female
Levels: male female
```

The terminology is that a factor has a set of levels, say k levels. Internally, a k-level factor consists of two items: a vector of integers between 1 and k and a character vector, of length k, containing strings with the corresponding labels.

Above, we created the factor from an integer vector; alternatively, it could have been constructed from other numerical, character, or logical vectors. Ordinal information may also be stored in a factor by setting the argument `ordered = TRUE` when calling `factor()`.

The advantage of this approach is that R knows when a certain variable is categorical and can choose appropriate methods automatically. For example, the labels can be used in printed output, different summary and plotting methods can be chosen, and contrast codings (e.g., dummy variables) can be computed in linear regressions. Note that for these actions the ordering of the levels can be important.

Missing values

Many data sets contain observations for which certain variables are unavailable. Econometric software needs ways to deal with this. In R, such missing values are coded as `NA` (for "not available"). All standard computations on `NA` become `NA`.

Special care is needed when reading data that use a different encoding. For example, when preparing the package **AER**, we encountered several data sets that employed -999 for missing values. If a file `mydata.txt` contains missing values coded in this way, they may be converted to `NA` using the argument `na.strings` when reading the file:

```
R> newdata <- read.table("mydata.txt", na.strings = "-999")
```

To query whether certain observations are `NA` or not, the function `is.na()` is provided.

2.6 Object Orientation

Somewhat vaguely, object-oriented programming (OOP) refers to a paradigm of programming where users/developers can create objects of a certain "class" (that are required to have a certain structure) and then apply "methods" for certain "generic functions" to these objects. A simple example in R is the function `summary()`, which is a generic function choosing, depending on the class of its argument, the summary method defined for this class. For example, for the numerical vector x and the factor g used above,

```
R> x <- c(1.8, 3.14, 4, 88.169, 13)
R> g <- factor(rep(c(0, 1), c(2, 4)), levels = c(0, 1),
+    labels = c("male", "female"))
```

the `summary()` call yields different types of results:

```
R> summary(x)
```

```
   Min. 1st Qu.  Median    Mean 3rd Qu.    Max.
   1.80    3.14    4.00   22.00   13.00   88.20
```

```
R> summary(g)
```

```
  male female
    2      4
```

For the numerical vector x, a five-number summary (i.e., the minimum and maximum, the median, and the first and third quartiles) along with the mean are reported, and for the factor g a simple frequency table is returned. This shows that R has different summary() methods available for different types of classes (in particular, it knows that a five-number summary is not sensible for categorical variables). In R, every object has a class that can be queried using the function class()

```
R> class(x)
```

```
[1] "numeric"
```

```
R> class(g)
```

```
[1] "factor"
```

which is used internally for calling the appropriate method for a generic function.

In fact, R offers several paradigms of object orientation. The base installation comes with two different OOP systems, usually called S3 (Chambers and Hastie 1992) and S4 (Chambers 1998). The S3 system is much simpler, using a dispatch mechanism based on a naming convention for methods. The S4 system is more sophisticated and closer to other OOP concepts used in computer science, but it also requires more discipline and experience. For most tasks, S3 is sufficient, and therefore it is the only OOP system (briefly) discussed here.

In S3, a generic function is defined as a function with a certain list of arguments and then a UseMethod() call with the name of the generic function. For example, printing the function summary() reveals its definition:

```
R> summary
```

```
function (object, ...)
UseMethod("summary")
<environment: namespace:base>
```

It takes a first required argument object plus an arbitrary number of further arguments passed through ... to its methods. What happens if this function is applied to an object, say of class "foo", is that R tries to apply the function summary.foo() if it exists. If not, it will call summary.default() if such a default method exists (which is the case for summary()). Furthermore, R objects can have a vector of classes (e.g., c("foo", "bar"), which means that such objects are of class "foo" inheriting from "bar"). In this case, R first tries to apply summary.foo(), then (if this does not exist) summary.bar(), and then (if both do not exist) summary.default(). All methods that are currently defined for a generic function can be queried using methods(); e.g., methods(summary) will return a (long) list of methods for all sorts of different classes. Among them is a method summary.factor(), which is used when summary(g) is called. However, there is no summary.numeric(), and hence summary(x) is handled by summary.default(). As it is not recommended to call methods directly, some methods are marked as being non-visible to the user, and these cannot (easily) be called directly. However, even for visible methods, we stress that in most situations it is clearly preferred to use, for example, summary(g) instead of summary.factor(g).

To illustrate how easy it is to define a class and some methods for it, let us consider a simple example. We create an object of class "normsample" that contains a sample from a normal distribution and then define a summary() method that reports the empirical mean and standard deviation for this sample. First, we write a simple class creator. In principle, it could have any name, but it is often called like the class itself:

```
R> normsample <- function(n, ...) {
+    rval <- rnorm(n, ...)
+    class(rval) <- "normsample"
+    return(rval)
+ }
```

This function takes a required argument n (the sample size) and further arguments ..., which are passed on to rnorm(), the function for generating normal random numbers. In addition to the sample size, it takes further arguments—the mean and the standard deviation; see ?rnorm. After generation of the vector of normal random numbers, it is assigned the class "normsample" and then returned.

```
R> set.seed(123)
R> x <- normsample(10, mean = 5)
R> class(x)
```

```
[1] "normsample"
```

To define a summary() method, we create a function summary.normsample() that conforms with the argument list of the generic (although ... is not used here) and computes the sample size, the empirical mean, and the standard deviation.

```
R> summary.normsample <- function(object, ...) {
+    rval <- c(length(object), mean(object), sd(object))
+    names(rval) <- c("sample size","mean","standard deviation")
+    return(rval)
+ }
```

Hence, calling

```
R> summary(x)
```

```
       sample size            mean standard deviation
          10.0000         5.0746              0.9538
```

automatically finds our new `summary()` method and yields the desired output.

Other generic functions with methods for most standard R classes are `print()`, `plot()`, and `str()`, which print, plot, and summarize the structure of an object, respectively.

2.7 R Graphics

It is no coincidence that early publications on S and R, such as Becker and Chambers (1984) and Ihaka and Gentleman (1996), are entitled "S: An Interactive Environment for Data Analysis and Graphics" and "R: A Language for Data Analysis and Graphics", respectively. R indeed has powerful graphics.

Here, we briefly introduce the "conventional" graphics as implemented in base R. R also comes with a new and even more flexible graphics engine, called **grid** (see Murrell 2005), that provides the basis for an R implementation of "trellis"-type graphics (Cleveland 1993) in the package **lattice** (Sarkar 2002), but these are beyond the scope of this book. An excellent overview of R graphics is given in Murrell (2005).

The function `plot()`

The basic function is the default `plot()` method. It is a generic function and has methods for many objects, including data frames, time series, and fitted linear models. Below, we describe the default `plot()` method, which can create various types of scatterplots, but many of the explanations extend to other methods as well as to other high-level plotting functions.

The scatterplot is probably the most common graphical display in statistics. A scatterplot of `y` vs. `x` is available using `plot(x, y)`. For illustration, we again use the `Journals` data from our package **AER**, taken from Stock and Watson (2007). As noted in Section 1.1, the data provide some information on subscriptions to economics journals at US libraries for the year 2000. The file contains 180 observations (the journals) on 10 variables, among them the number of library subscriptions (`subs`), the library subscription price (`price`), and the total number of citations for the journal (`citations`). These data will reappear in Chapter 3.

Here, we are interested in the relationship between the number of subscriptions and the price per citation. The following code chunk derives the required variable `citeprice` and plots the number of library subscriptions against it in logarithms:

```
R> data("Journals")
R> Journals$citeprice <- Journals$price/Journals$citations
R> attach(Journals)
```

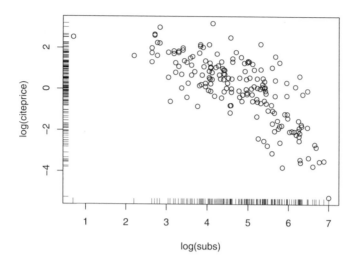

Fig. 2.2. Scatterplot of the journals data with ticks added.

```
R> plot(log(subs), log(citeprice))
R> rug(log(subs))
R> rug(log(citeprice), side = 2)
R> detach(Journals)
```

The function rug() adds ticks, thus visualizing the marginal distributions of
the variables, along one or both axes of an existing plot. Figure 2.2 has ticks
on both of the horizontal and vertical axes. An alternative way of specifying
plot(x, y) is to use the formula method of plot(); i.e., plot(y ~ x). This
leads to the same scatterplot but has the advantage that a data argument
can be specified. Hence we can avoid attaching and detaching the data frame:

```
R> plot(log(subs) ~ log(citeprice), data = Journals)
```

Graphical parameters

All this looks deceptively simple, but the result can be modified in numerous ways. For example, plot() takes a type argument that controls whether
points (type = "p", the default), lines (type = "l"), both (type = "b"),
stair steps (type = "s"), or further types of plots are generated. The annotation can be modified by changing the main title or the xlab and ylab axis
labels. See ?plot for more details.

Additionally, there are several dozen graphical parameters (see ?par for
the full list) that can be modified either by setting them with par() or by

Table 2.1. A selective list of arguments to `par()`.

Argument	Description
axes	should axes be drawn?
bg	background color
cex	size of a point or symbol
col	color
las	orientation of axis labels
lty, lwd	line type and line width
main, sub	title and subtitle
mar	size of margins
mfcol, mfrow	array defining layout for several graphs on a plot
pch	plotting symbol
type	types (see text)
xlab, ylab	axis labels
xlim, ylim	axis ranges
xlog, ylog, log	logarithmic scales

supplying them to the `plot()` function. We cannot explain all of these here, but we will highlight a few important parameters: `col` sets the color(s) and `xlim` and `ylim` adjust the plotting ranges. If points are plotted, `pch` can modify the plotting character and `cex` its character extension. If lines are plotted, `lty` and `lwd` specify the line type and width, respectively. The size of labels, axis ticks, etc., can be changed by further `cex`-type arguments such as `cex.lab` and `cex.axis`. A brief list of arguments to `par()` is provided in Table 2.1. This is just the tip of the iceberg, and further graphical parameters will be introduced as we proceed.

As a simple example, readers may want to try

```
R> plot(log(subs) ~ log(citeprice), data = Journals, pch = 20,
+     col = "blue", ylim = c(0, 8), xlim = c(-7, 4),
+     main = "Library subscriptions")
```

This yields solid circles (`pch = 20`) instead of the default open ones, drawn in blue, and there are wider ranges in the x and y directions; there is also a main title.

It is also possible to add further layers to a plot. Thus, `lines()`, `points()`, `text()`, and `legend()` add what their names suggest to an existing plot. For example, `text(-3.798, 5.846, "Econometrica", pos = 2)` puts a character string at the indicated location (i.e., to the left of the point). In regression analyses, one often wants to add a regression line to a scatterplot. As seen in Chapter 1, this is achieved using `abline(a, b)`, where a is the intercept and b is the slope.

At this point, there does not seem to be a great need for all this; however, most users require fine control of visual displays at some point, especially when publication-quality plots are needed. We refrain from presenting artificial examples toying with graphics options; instead we shall introduce variations of the standard displays as we proceed.

Of course, there are many further plotting functions besides the default `plot()` method. For example, standard statistical displays such as barplots, pie charts, boxplots, QQ plots, or histograms are available in the functions `barplot()`, `pie()`, `boxplot()`, `qqplot()`, and `hist()`. It is instructive to run `demo("graphics")` to obtain an overview of R's impressive graphics facilities.

Exporting graphics

In interactive use, graphics are typically written to a graphics window so that they can be inspected directly. However, after completing an analysis, we typically want to save the resulting graphics (e.g., for publication in a report, journal article, or thesis). For users of Microsoft Windows and Microsoft Word, a simple option is to "copy and paste" them into the Microsoft Word document. For other programs, such as LaTeX, it is preferable to export the graphic into an external file. For this, there exist various graphics devices to which plots can be written. Devices that are available on all platforms include the vector formats PostScript and PDF; other devices, such as the bitmap formats PNG and JPEG and the vector format WMF, are only available if supported by the system (see `?Devices` for further details). They all work in the same way: first the device is opened—e.g., the PDF device is opened by the function `pdf()`—then the commands creating the plot are executed, and finally the device is closed by `dev.off()`. A simple example creating a plot on a PDF device is:

```
R> pdf("myfile.pdf", height = 5, width = 6)
R> plot(1:20, pch = 1:20, col = 1:20, cex = 2)
R> dev.off()
```

This creates the PDF file `myfile.pdf` in the current working directory, which contains the graphic generated by the `plot()` call (see Figure 2.3). Incidentally, the plot illustrates a few of the parameters discussed above: it shows the first 20 plotting symbols (all shown in double size) and that in R a set of colors is also numbered. The first eight colors are black, red, green, blue, turquoise, violet, yellow, and gray. From color nine on, this vector is simply recycled.

Alternatively to opening, printing and closing a device, it is also possible to print an existing plot in the graphics window to a device using `dev.copy()` and `dev.print()`; see the corresponding manual page for more information.

Mathematical annotation of plots

A feature that particularly adds to R's strengths when it comes to publication-quality graphics is its ability to add mathematical annotation to plots (Murrell and Ihaka 2000). An S expression containing a mathematical expression can

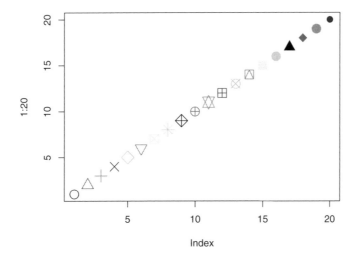

Fig. 2.3. Scatterplot written on a PDF device.

be passed to plotting functions without being evaluated; instead it is processed for annotation of the graph created. Readers familiar with LaTeX will have no difficulties in adapting to the syntax; for details, see `?plotmath` and `demo("plotmath")`. As an example, Figure 2.4 provides a plot of the density of the standard normal distribution (provided by `dnorm()` in R), including its mathematical definition

$$f(x) = \frac{1}{\sigma\sqrt{2\pi}} \, e^{-\frac{(x-\mu)^2}{2\sigma^2}} \, .$$

It is obtained via

```
R> curve(dnorm, from = -5, to = 5, col = "slategray", lwd = 3,
+    main = "Density of the standard normal distribution")
R> text(-5, 0.3, expression(f(x) == frac(1, sigma ~~
+    sqrt(2*pi)) ~~ e^{-frac((x - mu)^2, 2*sigma^2)}), adj = 0)
```

The function `curve()` plots the density function `dnorm()`, and then `text()` is used to add the `expression()` containing the formula to the plot. This example concludes our brief introduction to R graphics.

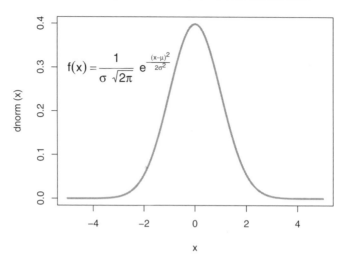

Fig. 2.4. Plot of the density of the standard normal distribution, including its mathematical expression.

2.8 Exploratory Data Analysis with R

In this section, we shall briefly illustrate some standard exploratory data analysis techniques. Readers seeking a more detailed introduction to basic statistics using R are referred to Dalgaard (2002).

We reconsider the CPS1985 data taken from Berndt (1991), which were encountered in our introductory R session when illustrating several regression methods. After making the data available via data(), some basic information can be queried by str():

```
R> data("CPS1985")
R> str(CPS1985)
```

```
'data.frame':        533 obs. of   11 variables:
 $ wage       : num    4.95  6.67  4.00  7.50 13.07 ...
 $ education  : int   9 12 12 12 13 10 12 16 12 12 ...
 $ experience : int   42 1 4 17 9 27 9 11 9 17 ...
 $ age        : int   57 19 22 35 28 43 27 33 27 35 ...
 $ ethnicity  : Factor w/ 3 levels "cauc","hispanic",..: 1 1 1..
 $ region     : Factor w/ 2 levels "south","other": 2 2 2 2 2 ..
 $ gender     : Factor w/ 2 levels "male","female": 2 1 1 1 1 ..
 $ occupation : Factor w/ 6 levels "worker","technical",..: 1 ..
 $ sector     : Factor w/ 3 levels "manufacturing",..: 1 1 3 3..
```

```
$ union     : Factor w/ 2 levels "no","yes": 1 1 1 2 1 1 1..
$ married   : Factor w/ 2 levels "no","yes": 2 1 1 2 1 1 1 2..
```

This reveals that this "data.frame" object comprises 533 observations on 11 variables, including the numerical variable wage, the integer variables education, experience, and age, and seven factors, each comprising two to six levels.

Instead of using the list-type view that str() provides, it is often useful to inspect the top (or the bottom) of a data frame in its rectangular representation. For this purpose, there exist the convenience functions head() and tail(), returning (by default) the first and last six rows, respectively. Thus

```
R> head(CPS1985)
```

```
  wage education experience age ethnicity region gender
1  4.95         9         42  57      cauc  other female
2  6.67        12          1  19      cauc  other   male
3  4.00        12          4  22      cauc  other   male
4  7.50        12         17  35      cauc  other   male
5 13.07        13          9  28      cauc  other   male
6  4.45        10         27  43      cauc  south   male
  occupation        sector union married
1     worker manufacturing    no     yes
2     worker manufacturing    no      no
3     worker         other    no      no
4     worker         other    no     yes
5     worker         other   yes      no
6     worker         other    no      no
```

Another useful way of gaining a quick overview of a data set is to use the summary() method for data frames, which provides a summary for each of the variables. As the type of the summary depends on the class of the respective variable, we inspect the summary() methods separately for various variables from CPS1985 below. Hence, the output of summary(CPS1985) is omitted here.

As the CPS1985 data are employed repeatedly in the following, we avoid lengthy commands such as CPS1985$education by attaching the data set. Also, to compactify subsequent output, we abbreviate two levels of occupation from "technical" to "techn" and from "management" to "mgmt".

```
R> levels(CPS1985$occupation)[c(2, 6)] <- c("techn", "mgmt")
R> attach(CPS1985)
```

Now variables are accessible by their names.

We proceed by illustrating exploratory analysis of single as well as pairs of variables, distinguishing among numerical variables, factors, and combinations thereof. We begin with the simplest kind, a single numerical variable.

One numerical variable

We will first look at the distribution of wages in the sample:

```
R> summary(wage)
```

```
   Min. 1st Qu.  Median    Mean 3rd Qu.    Max.
   1.00    5.25    7.78    9.03   11.20   44.50
```

This provides Tukey's five-number summary plus the mean wage. The mean and median could also have been obtained using

```
R> mean(wage)
```

```
[1] 9.031
```

```
R> median(wage)
```

```
[1] 7.78
```

and `fivenum()` computes the five-number summary. Similarly, `min()` and `max()` would have yielded the minimum and the maximum. Arbitrary quantiles can be computed by `quantile()`.

For measures of spread, there exist the functions

```
R> var(wage)
```

```
[1] 26.43
```

```
R> sd(wage)
```

```
[1] 5.141
```

returning the variance and the standard deviation, respectively.

Graphical summaries are also helpful. For numerical variables such as `wage`, density visualizations (via histograms or kernel smoothing) and boxplots are suitable. Boxplots will be considered below in connection with two-variable displays. Figure 2.5, obtained via

```
R> hist(wage, freq = FALSE)
R> hist(log(wage), freq = FALSE)
R> lines(density(log(wage)), col = 4)
```

shows the densities of `wage` and its logarithm (that is, areas under curves equal 1, resulting from `freq = FALSE`; otherwise absolute frequencies would have been depicted). Further arguments allow for fine tuning of the selection of the breaks in the histogram. Added to the histogram in the right panel is a kernel density estimate obtained using `density()`. Clearly, the distribution of the logarithms is less skewed than that of the raw data. Note that `density()` only computes the density coordinates and does not provide a plot; hence the estimate is added via `lines()`.

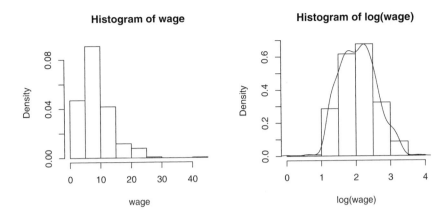

Fig. 2.5. Histograms of wages (left panel) and their logarithms with superimposed density (right panel).

One categorical variable

For categorical data, it makes no sense to compute means and variances; instead one needs a table indicating the frequencies with which the categories occur. If R is told that a certain variable is categorical (by making it a "factor"), it automatically chooses an appropriate summary:

```
R> summary(occupation)
```

worker	techn	services	office	sales	mgmt
155	105	83	97	38	55

This could also have been computed by table(occupation). If relative instead of absolute frequencies are desired, there exists the function prop.table():

```
R> tab <- table(occupation)
R> prop.table(tab)
```

```
occupation
```

worker	techn	services	office	sales	mgmt
0.2908	0.1970	0.1557	0.1820	0.0713	0.1032

Categorical variables are typically best visualized by barplots. If majorities are to be brought out, pie charts might also be useful. Thus

```
R> barplot(tab)
R> pie(tab)
```

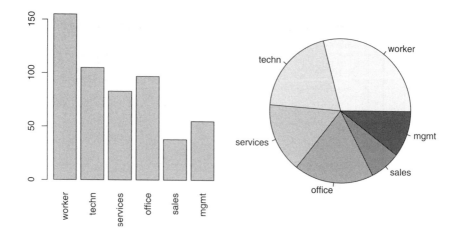

Fig. 2.6. Bar plot and pie chart of occupation.

provides Figure 2.6. Note that both functions expect the tabulated frequencies as input. In addition, calling plot(occupation) is equivalent to barplot(table(occupation)).

Two categorical variables

The relationship between two categorical variables is typically summarized by a contingency table. This can be created either by xtabs(), a function with a formula interface, or by table(), a function taking an arbitrary number of variables for cross-tabulation (and not only a single one as shown above).

We consider the factors occupation and gender for illustration:

```
R> xtabs(~ gender + occupation, data = CPS1985)
```

```
        occupation
gender  worker techn services office sales mgmt
  male     126    53       34     21    21   34
  female    29    52       49     76    17   21
```

which can equivalently be created by table(gender, occupation). A simple visualization is a mosaic plot (Hartigan and Kleiner 1981; Friendly 1994), which can be seen as a generalization of stacked barplots. The plot given in Figure 2.7 (also known as a "spine plot", a certain variant of the standard mosaic display), obtained via

```
R> plot(gender ~ occupation, data = CPS1985)
```

shows that the proportion of males and females changes considerably over the levels of occupation. In addition to the shading that brings out the

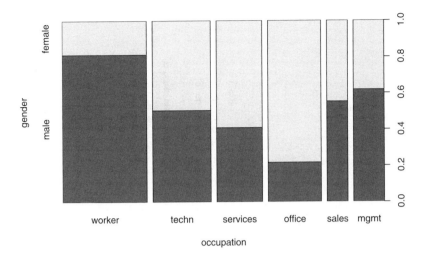

Fig. 2.7. Mosaic plot (spine plot) of gender versus occupation.

conditional distribution of **gender** given **occupation**, the widths of the bars visualize the marginal distribution of **occupation**, indicating that there are comparatively many workers and few salespersons.

Two numerical variables

We exemplify the exploratory analysis of the relationship between two numerical variables by using **wage** and **education**.

A summary measure for two numerical variables is the correlation coefficient, implemented in the function **cor()**. However, the standard (Pearson) correlation coefficient is not necessarily meaningful for positive and heavily skewed variables such as **wage**. We therefore also compute a nonparametric variant, Spearman's ϱ, which is available in **cor()** as an option.

```
R> cor(log(wage), education)
```

```
[1] 0.379
```

```
R> cor(log(wage), education, method = "spearman")
```

```
[1] 0.3798
```

Both measures are virtually identical and indicate only a modest amount of correlation here, see also the corresponding scatterplot in Figure 2.8:

```
R> plot(log(wage) ~ education)
```

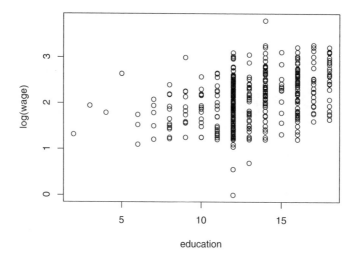

Fig. 2.8. Scatterplot of wages (in logs) versus education.

One numerical and one categorical variable

It is common to have both numerical and categorical variables in a data frame. For example, here we have `wage` and `gender`, and there might be some interest in the distribution of `wage` by `gender`. A suitable function for numerical summaries is `tapply()`. It applies, for a numerical variable as its first argument and a (list of) categorical variable(s) as its second argument, the function specified as the third argument. Hence, mean wages conditional on gender are available using

```
R> tapply(log(wage), gender, mean)
```

```
  male female
 2.165  1.935
```

Using similar commands, further descriptive measures or even entire summaries (just replace `mean` by `summary`) may be computed.

Suitable graphical displays are parallel boxplots and quantile-quantile (QQ) plots, as depicted in Figure 2.9. Recall that a boxplot (or "box-and-whiskers plot") is a coarse graphical summary of an empirical distribution. The box indicates "hinges" (approximately the lower and upper quartiles) and the median. The "whiskers" (lines) indicate the largest and smallest observations falling within a distance of 1.5 times the box size from the nearest hinge. Any observations falling outside this range are shown separately and would

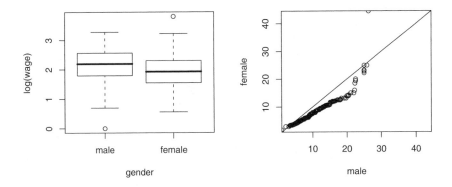

Fig. 2.9. Boxplot and QQ plot of wages stratified by gender.

be considered extreme or outlying (in an approximately normal sample). Note that there are several variants of boxplots in the literature.

The commands `plot(y ~ x)` and `boxplot(y ~ x)` both yield the same parallel boxplot if x is a "`factor`"; thus

```
R> plot(log(wage) ~ gender)
```

gives the left panel of Figure 2.9. It shows that the overall shapes of both distributions are quite similar and that males enjoy a substantial advantage, especially in the medium range. The latter feature is also brought out by the QQ plot (right panel) resulting from

```
R> mwage <- subset(CPS1985, gender == "male")$wage
R> fwage <- subset(CPS1985, gender == "female")$wage
R> qqplot(mwage, fwage, xlim = range(wage), ylim = range(wage),
+    xaxs = "i", yaxs = "i", xlab = "male", ylab = "female")
R> abline(0, 1)
```

where almost all points are below the diagonal (corresponding to identical distributions in both samples). This indicates that, for most quantiles, male wages are typically higher than female wages.

We end this section by detaching the data:

```
R> detach(CPS1985)
```

2.9 Exercises

1. Create a square matrix, say A, with entries $a_{ii} = 2$, $i = 2, \ldots, n - 1$, $a_{11} = a_{nn} = 1$, $a_{i,i+1} = a_{i,i-1} = -1$, and $a_{ij} = 0$ elsewhere. (Where does this matrix occur in econometrics?)

2. "PARADE" is the Sunday newspaper magazine supplementing the Sunday or weekend edition of some 500 daily newspapers in the United States of America. An important yearly feature is an article providing information on some 120–150 "randomly" selected US citizens, indicating their profession, hometown and state, and their yearly earnings. The Parade2005 data contain the 2005 version, amended by a variable indicating celebrity status (motivated by substantial oversampling of celebrities in these data). For the Parade2005 data:

   (a) Determine the mean earnings in California. Explain the result.
   (b) Determine the number of individuals residing in Idaho. (What does this say about the data set?)
   (c) Determine the mean and the median earnings of celebrities. Comment.
   (d) Obtain boxplots of log(earnings) stratified by celebrity. Comment.

3. For the Parade2005 data of the preceding exercise, obtain a kernel density estimate of the earnings for the full data set. It will be necessary to transform the data to logarithms (why?). Comment on the result. Be sure to try out some arguments to density(), in particular the plug-in bandwidth bw.

4. Consider the CPS1988 data, taken from Bierens and Ginther (2001). (These data will be used for estimating an earnings equation in the next chapter.)

   (a) Obtain scatterplots of the logarithm of the real wage (wage) versus experience and versus education.
   (b) In fact, education corresponds to years of schooling and therefore takes on only a limited number of values. Transform education into a factor and obtain parallel boxplots of wage stratified by the levels of education. Repeat for experience.
   (c) The data set contains four additional factors, ethnicity, smsa, region, and parttime. Obtain suitable graphical displays of log(wage) versus each of these factors.

3

Linear Regression

The linear regression model, typically estimated by ordinary least squares (OLS), is the workhorse of applied econometrics. The model is

$$y_i = x_i^\top \beta + \varepsilon_i, \quad i = 1, \ldots, n,$$

or, in matrix form,

$$y = X\beta + \varepsilon,$$

where y is an $n \times 1$ vector containing the dependent variable, x_i is the (column) vector of covariates for observation i—thus $X = (x_1, \ldots, x_n)^\top$ is the $n \times k$ regressor matrix, or model matrix (whose columns contain the regressors)—and β is a $k \times 1$ vector of regression coefficients. Furthermore, ε is the $n \times 1$ vector of disturbances (or error terms). Assumptions on the error terms depend on the context. For cross sections, $\mathsf{E}(\varepsilon|X) = 0$ (exogeneity) and $\mathsf{Var}(\varepsilon|X) = \sigma^2 I$ (conditional homoskedasticity and lack of correlation) are common. However, for time series data, exogeneity is too strong an assumption, and it is commonly replaced by predeterminedness; i.e., $\mathsf{E}(\varepsilon_j|x_i) = 0$, $i \le j$. Methods for checking these assumptions are discussed in Chapter 4.

We assume that readers are familiar with the basics of the linear regression model, say at the level of Baltagi (2002) or Greene (2003). To fix notation, let $\hat{\beta} = (X^\top X)^{-1} X^\top y$ denote the familiar OLS estimator of β. The corresponding fitted values are $\hat{y} = X\hat{\beta}$, the residuals are $\hat{\varepsilon} = y - \hat{y}$, and the residual sum of squares (RSS) is $\hat{\varepsilon}^\top \hat{\varepsilon}$.

In R, models are typically fitted by calling a model-fitting function, in this case `lm()`, with a "formula" object describing the model and a "data.frame" object containing the variables used in the formula. Most fitting functions, including `lm()`, take further arguments, providing a more detailed description of the model or control parameters for the fitting algorithm. By subsuming such further arguments in ..., a prototypical call looks like

```
fm <- lm(formula, data, ...)
```

C. Kleiber, A. Zeileis, *Applied Econometrics with R,*
DOI: 10.1007/978-0-387-77318-6_3, © Springer Science+Business Media, LLC 2008

returning a fitted-model object, here stored in `fm`. This fitted model can subsequently be printed, summarized, or visualized; fitted values and residuals can be extracted or predictions on new data computed. Methods for suitable generic functions such as `summary()`, `residuals()`, or `predict()`, etc., are available for all standard models.

Many models can be seen as extensions of the linear regression model. Analogously, the R function `lm()` for fitting linear models is a prototypical fitting function–many other model fitters have similar interfaces and can be used in virtually the same way. Hence, a good grasp of `lm()` and the associated methods is essential for all subsequent chapters. Thus, this chapter begins with simple linear regression to introduce the main fitting function and associated generic functions. This is followed by multiple regression and partially linear models. Some fine points of the formula notation, mainly in connection with factors and interactions, are discussed in a starred section; it may be skipped at first reading. The remaining sections consider special types of data, namely time series and panel data, and also systems of linear equations.

3.1 Simple Linear Regression

We begin with a small example to provide a feel for the process. The data set `Journals` is taken from Stock and Watson (2007). As noted in the preceding chapters, the data provide some information on subscriptions to economics journals at US libraries for the year 2000. Bergstrom (2001) argues that commercial publishers are charging excessive prices for academic journals and also suggests ways that economists can deal with this problem. We refer the interested reader to Bergstrom's journal pricing page, currently at `http://www.econ.ucsb.edu/~tedb/Journals/jpricing.html`, for further information.

The `Journals` data frame contains 180 observations (the journals) on 10 variables, among them the number of library subscriptions (`subs`), the library subscription price (`price`), and the total number of citations for the journal (`citations`). Here, we require only some of the variables, and hence we first generate a smaller data frame for compactness. It contains the relevant variables as well as a transformed variable, the price per citation. The data can be loaded, transformed, and summarized via

```
R> data("Journals")
R> journals <- Journals[, c("subs", "price")]
R> journals$citeprice <- Journals$price/Journals$citations
R> summary(journals)
```

```
      subs             price           citeprice
 Min.   :   2    Min.   :  20    Min.   : 0.00522
 1st Qu.:  52    1st Qu.: 134    1st Qu.: 0.46450
 Median : 122    Median : 282    Median : 1.32051
```

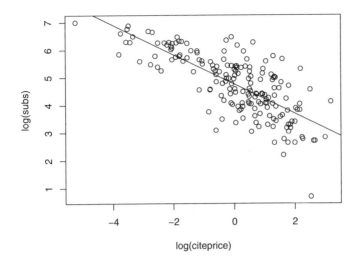

Fig. 3.1. Scatterplot of the journals data with least-squares line.

Mean	: 197	Mean	: 418	Mean	: 2.54845
3rd Qu.	: 268	3rd Qu.	: 541	3rd Qu.	: 3.44017
Max.	:1098	Max.	:2120	Max.	:24.45946

In view of the wide range of the variables, combined with a considerable amount of skewness, it is useful to take logarithms.

The goal is to estimate the effect of the price per citation on the number of library subscriptions. To explore this issue quantitatively, we will fit a linear regression model,

$$\texttt{log(subs)}_i = \beta_1 + \beta_2 \, \texttt{log(citeprice)}_i + \varepsilon_i.$$

The primary function for this is `lm()`, which estimates a linear regression using ordinary least squares (OLS). As motivated above, a linear model is fitted by supplying a formula that describes the model plus a data frame.

Here, the formula of interest is `log(subs) ~ log(citeprice)`; i.e., `log(subs)` explained by `log(citeprice)`. This can be used both for plotting and for model fitting:

```
R> plot(log(subs) ~ log(citeprice), data = journals)
R> jour_lm <- lm(log(subs) ~ log(citeprice), data = journals)
R> abline(jour_lm)
```

The resulting plot is shown in Figure 3.1. `abline()` extracts the coefficients of the fitted model and adds the corresponding regression line to the plot.

The right-hand side (RHS) of the formula needs only to specify the variable `log(citeprice)` since a constant term is included by default. It may be added explicitly using `1 + log(citeprice)`, which, however, does not make a difference. Recall that operators such as `+` on the RHS have the special meaning of the Wilkinson-Rogers notation (see Section 2.4), *not* their arithmetical meaning. Further details on formulas in fitting functions are provided in the following sections. If the argument `data` is specified, it gives a data frame from which variables are selected ahead of the search path. We recommend to always use this argument.

The function `lm()` returns a fitted-model object, here stored as `jour_lm`. It is an object of class "lm"

```
R> class(jour_lm)
```

```
[1] "lm"
```

which is essentially a list; the names of its components can be queried via

```
R> names(jour_lm)
```

```
 [1] "coefficients"  "residuals"       "effects"
 [4] "rank"          "fitted.values"   "assign"
 [7] "qr"            "df.residual"     "xlevels"
[10] "call"          "terms"           "model"
```

Thus, `jour_lm` consists of 12 components, among them `coefficients`, `residuals`, and `call` (the function call), plus various quantities computed when fitting the model, such as the rank of the model matrix. Typing `str(jour_lm)` provides a more detailed view. As for any other list, all components are directly accessible using, for example, `jour_lm$rank`. However, for most tasks, there is no need to do so since generic functions are available that extract quantities of interest. An overview of generics that have methods for "lm" objects is given in Table 3.1.

The output from `summary()` is self-explanatory. For our model, we obtain

```
R> summary(jour_lm)
```

```
Call:
lm(formula = log(subs) ~ log(citeprice), data = journals)

Residuals:
    Min      1Q  Median      3Q     Max
-2.7248 -0.5361  0.0372  0.4662  1.8481

Coefficients:
               Estimate Std. Error t value Pr(>|t|)
(Intercept)      4.7662     0.0559    85.2   <2e-16
log(citeprice)  -0.5331     0.0356   -15.0   <2e-16
```

Table 3.1. Generic functions for fitted (linear) model objects.

Function	Description
`print()`	simple printed display
`summary()`	standard regression output
`coef()`	(or `coefficients()`) extracting the regression coefficients
`residuals()`	(or `resid()`) extracting residuals
`fitted()`	(or `fitted.values()`) extracting fitted values
`anova()`	comparison of nested models
`predict()`	predictions for new data
`plot()`	diagnostic plots
`confint()`	confidence intervals for the regression coefficients
`deviance()`	residual sum of squares
`vcov()`	(estimated) variance-covariance matrix
`logLik()`	log-likelihood (assuming normally distributed errors)
`AIC()`	information criteria including AIC, BIC/SBC (assuming normally distributed errors)

```
Residual standard error: 0.75 on 178 degrees of freedom
Multiple R-squared: 0.557,          Adjusted R-squared: 0.555
F-statistic:  224 on 1 and 178 DF,  p-value: <2e-16
```

This gives a brief numerical summary of the residuals as well as a table of the estimated regression coefficients along with their standard errors. Here, we obtain that the intercept of the fitted line is 4.7662, with a standard error of 0.0559, and the estimated slope is -0.5331, with a standard error of 0.0356. Also included are t statistics (the ratio of the coefficient estimate and its standard error) and p values corresponding to individual tests of the hypothesis "the true coefficient equals 0". Here, both p values are tiny, indicating that the regressor explains a substantial part of the variation in the data and that the intercept is significantly different from zero, at any resonable level. Also, two versions of R^2—the standard multiple R^2 and Theil's adjusted R^2—tell us that the model explains more than 50% of the variation in the data, a reasonable amount for cross sections. Finally, the F statistic corresponds to an F test of the hypothesis that all regressors (excluding the intercept term) are jointly significant. Here, with just a single regressor, the p value is of course identical to that of the t test for the coefficient of `log(citeprice)`, the F statistic being the squared value of the t statistic. Both indicate that the regressor `log(citeprice)` is highly significant.

It is instructive to take a brief look at what the `summary()` method returns for a fitted "lm" object:

```
R> jour_slm <- summary(jour_lm)
R> class(jour_slm)
```

```
[1] "summary.lm"
```

```
R> names(jour_slm)
```

```
 [1] "call"          "terms"        "residuals"
 [4] "coefficients"  "aliased"      "sigma"
 [7] "df"            "r.squared"    "adj.r.squared"
[10] "fstatistic"    "cov.unscaled"
```

This indicates that the "summary.lm" object jour_slm is a list whose components are quite similar to those of "lm" objects but now contain the summary information; e.g., the adjusted R^2 in jour_slm$adj.r.squared or the full matrix of coefficients, standard errors, t statistics, and p values in

```
R> jour_slm$coefficients
```

```
                Estimate Std. Error t value  Pr(>|t|)
(Intercept)       4.7662    0.05591   85.25 2.954e-146
log(citeprice)   -0.5331    0.03561  -14.97   2.564e-33
```

Again, a more complete overview of the information stored in jour_slm may be obtained by calling str(jour_slm).

Analysis of variance

Some of the information provided in the summary is also available using different extractor functions. For example, the information appearing at the bottom can also be summarized in the form of an analysis of variance (ANOVA) table:

```
R> anova(jour_lm)
```

```
Analysis of Variance Table
```

```
Response: log(subs)
                Df Sum Sq Mean Sq F value Pr(>F)
log(citeprice)   1  125.9   125.9     224 <2e-16
Residuals      178  100.1     0.6
```

The ANOVA table breaks the sum of squares about the mean (for the dependent variable, here log(subs)) into two parts: a part that is accounted for by a linear function of log(citeprice) and a part attributed to residual variation. The total sum of squares (about the mean of log(subs)) is 225.99 (= 125.93 + 100.06). Including the regressor log(citeprice) reduced this by 125.93, yielding a residual sum of squares (RSS) equal to 100.06. For comparing the reduction with the RSS, it is best to look at the column Mean Sq. The mean square of the reduction was 125.93 (equal to the raw sum of squares since there is only one regressor), yielding 0.56 as the mean square of the residual. The ratio of these quantities is 224.0369, the value of the F statistic for testing the hypothesis $H_0 : \beta_2 = 0$. The associated p value is tiny, again indicating that the regressor log(citeprice) is highly significant. Of course, this is exactly the F test given in the summary above, which in turn is equivalent to the t test, as noted before. The column Df provides degrees of freedom (df): 1 df is attributed to the regressor log(citeprice), while $180 - 2 = 178$ are attributed to the residuals (two regression coefficients were estimated).

The anova() method not only can produce an ANOVA table for a single "lm" object, but can also be used to compare several nested "lm" models using F tests. This will be illustrated in the next section.

Point and interval estimates

To extract the estimated regression coefficients $\hat{\beta}$, the function coef() can be used:

```
R> coef(jour_lm)
```

```
(Intercept) log(citeprice)
     4.7662        -0.5331
```

It is good practice to give a measure of error along with every estimate. One way to do this is to provide a confidence interval. This is available via the extractor function confint(); thus

```
R> confint(jour_lm, level = 0.95)
```

```
                 2.5 %   97.5 %
(Intercept)     4.6559   4.8765
log(citeprice) -0.6033  -0.4628
```

The default level is 95%, so specification of level = 0.95 was not really needed here.

Prediction

Often a regression model is used for prediction. Recall that there are two types of predictions: the prediction of points on the regression line and the prediction of a new data value. The standard errors of predictions for new data take into account both the uncertainty in the regression line and the variation of the individual points about the line. Thus, the prediction interval for prediction of new data is larger than that for prediction of points on the line. The function predict() provides both types of standard errors.

For the journals data, we might be interested in the expected number of subscriptions for a journal whose price per citation equals 2.11 (this roughly corresponds to the value for the *Journal of Applied Econometrics*, a journal that is owned by a commercial publisher and fairly expensive) or in the number of library subscriptions itself. Both types of prediction intervals are given by the following code:

```
R> predict(jour_lm, newdata = data.frame(citeprice = 2.11),
+    interval = "confidence")

    fit    lwr    upr
1 4.368 4.247 4.489

R> predict(jour_lm, newdata = data.frame(citeprice = 2.11),
+    interval = "prediction")

    fit    lwr    upr
1 4.368 2.884 5.853
```

Of course, the point estimates (denoted fit) are identical; only the intervals differ. Recall that these intervals are based on the t distribution with 178 df (residual df in our case) and are exact under the assumption of (conditionally) normally distributed disturbances. By default, no intervals are computed and the data used for fitting are also used as newdata, so that both predict(jour_lm) and fitted(jour_lm) compute the fitted values $\hat{y}$.

The prediction intervals can also be used for computing and visualizing confidence bands. In the following code chunk, we set up an auxiliary variable lciteprice with equidistant points on the scale of log(citeprice), then compute the corresponding prediction intervals, and finally visualize them using plot() and lines(), which here adds the lines for the fitted values and the confidence bands. The resulting plot is shown in Figure 3.2.

```
R> lciteprice <- seq(from = -6, to = 4, by = 0.25)
R> jour_pred <- predict(jour_lm, interval = "prediction",
+    newdata = data.frame(citeprice = exp(lciteprice)))
R> plot(log(subs) ~ log(citeprice), data = journals)
R> lines(jour_pred[, 1] ~ lciteprice, col = 1)
R> lines(jour_pred[, 2] ~ lciteprice, col = 1, lty = 2)
R> lines(jour_pred[, 3] ~ lciteprice, col = 1, lty = 2)
```

Plotting "lm" objects

The plot() method for class lm() provides six types of diagnostic plots, four of which are shown by default. Figure 3.3 depicts the result for the journals regression. We set the graphical parameter mfrow to c(2, 2) using the par() function, creating a 2 × 2 matrix of plotting areas to see all four plots simultaneously:

```
R> par(mfrow = c(2, 2))
R> plot(jour_lm)
R> par(mfrow = c(1, 1))
```

The first provides a graph of residuals versus fitted values, perhaps the most familiar of all diagnostic plots for residual analysis. The second is a QQ plot for normality. Its curvature (or rather its lack thereof) shows that the residuals

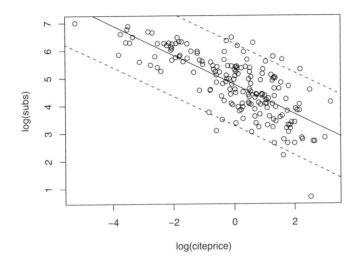

Fig. 3.2. Scatterplot with prediction intervals for the journals data.

more or less conform with normality. Plots three and four are a scale-location plot and a plot of standardized residuals against leverages, respectively. Their discussion will be postponed to Chapter 4. For the moment, we just note that `plot(jour_lm)` already provides some indication of the fit of the model.

In our case, there appear to be difficulties with observations "MEPiTE", "RoRPE", "IO", "BoIES", and "Ecnmt", corresponding to the journals *MOCT-MOST: Economic Policy in Transitional Economics*, *Review of Radical Political Economics*, *International Organization*, *Bulletin of Indonesian Economic Studies*, and *Econometrica*, each of which is singled out in at least one of the four plots. A closer look reveals that all these journals are not overly expensive, they are unusual in that they are either heavily cited (*Econometrica*), resulting in a low price per citation, or have only a few citations, resulting in a rather high price per citation. Incidentally, all plots may be accessed individually using, for example, `plot(jour_lm, which = 2)`, if only the QQ plot is desired.

Testing a linear hypothesis

The standard regression output as provided by `summary()` only indicates individual significance of each regressor and joint significance of all regressors in the form of t and F statistics, respectively. Often it is necessary to test more general hypotheses. This is possible using the function `linear.hypothesis()` from the **car** package, the package accompanying Fox (2002). (**car** will

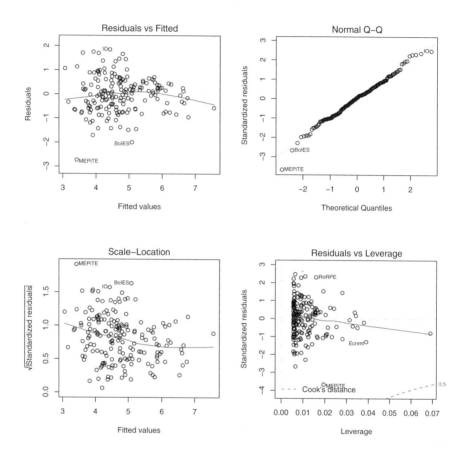

Fig. 3.3. Diagnostic plots for the journals data.

automatically be installed when installing **AER**, and it will also automatically be loaded when loading **AER**.) Recall that a linear hypothesis is of the general form

$$R\beta = r, \qquad (3.1)$$

where β is the $k \times 1$ vector of regression coefficients, R is a $q \times k$ matrix, and r is a $q \times 1$ vector. (In this section, $k = 2$.)

Suppose we want to test, for the `journals` data, the hypothesis that the elasticity of the number of library subscriptions with respect to the price per citation equals -0.5. Since this corresponds to the linear hypothesis $H_0 : \beta_2 = -0.5$, we may proceed as follows: `linear.hypothesis()` requires a fitted-model object and a specification of the linear hypothesis from (3.1). This can be simply specified in a character vector:

```
R> linear.hypothesis(jour_lm, "log(citeprice) = -0.5")

Linear hypothesis test

Hypothesis:
log(citeprice) = -0.5

Model 1: log(subs) ~ log(citeprice)
Model 2: restricted model

  Res.Df   RSS  Df Sum of Sq     F Pr(>F)
1    178 100.1
2    179 100.5  -1      -0.5  0.86   0.35
```

Alternatively, the hypothesis.matrix R and the rhs vector r could be specified explicitly via linear.hypothesis(jour_lm, hypothesis.matrix = c(0, 1), rhs = -0.5), leading to equivalent results. The output of linear.hypothesis() is similar to that of anova(): it gives the models along with their degrees of freedom (Df), RSS, and associated F statistic, here suggesting that the elasticity of interest is not substantially different from −0.5.

3.2 Multiple Linear Regression

In economics, most regression analyses comprise more than a single regressor. Often there are regressors of a special type, usually referred to as "dummy variables" in econometrics, which are used for coding categorical variables. Furthermore, it is often also necessary to transform regressors or dependent variables. To illustrate how to deal with this in R, we consider a standard task in labor economics, estimation of a wage equation in semilogarithmic form. Here, we employ the CPS1988 data frame collected in the March 1988 Current Population Survey (CPS) by the US Census Bureau and analyzed by Bierens and Ginther (2001). One reason for choosing this data set, comprising 28,155 observations, was to provide an "industry-strength" example instead of the common textbook examples containing at most a few hundred observations. These are cross-section data on males aged 18 to 70 with positive annual income greater than US\$ 50 in 1992 who are not self-employed or working without pay. Wages are deflated by the deflator of personal consumption expenditures for 1992. A summary of the data set can be obtained as usual:

```
R> data("CPS1988")
R> summary(CPS1988)

      wage           education       experience     ethnicity
 Min.   :   50   Min.   : 0.0   Min.   :-4.0   cauc:25923
```

```
1st Qu.:   309    1st Qu.:12.0    1st Qu.: 8.0    afam: 2232
Median :   522    Median :12.0    Median :16.0
Mean   :   604    Mean   :13.1    Mean   :18.2
3rd Qu.:   783    3rd Qu.:15.0    3rd Qu.:27.0
Max.   :18777    Max.   :18.0    Max.   :63.0
  smsa               region       parttime
no : 7223    northeast:6441    no :25631
yes:20932    midwest  :6863    yes: 2524
             south    :8760
             west     :6091
```

Here, wage is the wage in dollars per week, education and experience are measured in years, and ethnicity is a factor with levels Caucasian ("cauc") and African-American ("afam"). There are three further factors, smsa, region, and parttime, indicating residence in a standard metropolitan statistical area (SMSA), the region within the United States of America, and whether the individual works part-time. Note that the CPS does not provide actual work experience. It is therefore customary to compute experience as age - education - 6 (as was done by Bierens and Ginther); this may be considered potential experience. This quantity may become negative, which explains the 438 observations with this property in the CPS1988 data.

The model of interest is

$$\log(\text{wage}) = \beta_1 + \beta_2 \, \text{experience} + \beta_3 \, \text{experience}^2$$
$$+\beta_4 \, \text{education} + \beta_5 \, \text{ethnicity} + \varepsilon \qquad (3.2)$$

which can be fitted in R using

```
R> cps_lm <- lm(log(wage) ~ experience + I(experience^2) +
+    education + ethnicity, data = CPS1988)
```

The formula in the lm() call takes into account the semilogarithmic form and also specifies the squared regressor experience^2. It has to be insulated by I() so that the operator ^ has its original arithmetic meaning (and not its meaning as a formula operator for specifying interactions; see below). The summary

```
R> summary(cps_lm)

Call:
lm(formula = log(wage) ~ experience + I(experience^2) +
    education + ethnicity, data = CPS1988)

Residuals:
   Min     1Q Median     3Q    Max
-2.943 -0.316  0.058  0.376  4.383
```

```
Coefficients:
                Estimate Std. Error t value Pr(>|t|)
(Intercept)     4.321395   0.019174   225.4   <2e-16
experience      0.077473   0.000880    88.0   <2e-16
I(experience^2) -0.001316  0.000019   -69.3   <2e-16
education       0.085673   0.001272    67.3   <2e-16
ethnicityafam  -0.243364   0.012918   -18.8   <2e-16

Residual standard error: 0.584 on 28150 degrees of freedom
Multiple R-squared: 0.335,        Adjusted R-squared: 0.335
F-statistic: 3.54e+03 on 4 and 28150 DF,  p-value: <2e-16
```

reveals that all coefficients have the expected sign, and the corresponding variables are highly significant (not surprising in a sample as large as the present one). Specifically, according to this specification, the return on education is 8.57% per year.

Dummy variables and contrast codings

Note that the level "cauc" of ethnicity does not occur in the output, as it is taken as the reference category. Hence, there is only one ethnicity effect, which gives the difference in intercepts between the "afam" and the "cauc" groups. In statistical terminology, this is called a "treatment contrast" (where the "treatment" "afam" is compared with the reference group "cauc") and corresponds to what is called a "dummy variable" (or "indicator variable") for the level "afam" in econometric jargon.

In R, (unordered) factors are automatically handled like this when they are included in a regression model. Internally, R produces a dummy variable for each level of a factor and resolves the resulting overspecification of the model (if an intercept or another factor is included in the model) by applying "contrasts"; i.e., a constraint on the underlying parameter vector. Contrasts are attributed to each factor and can be queried and changed by contrasts(). The default for unordered factors is to use all dummy variables except the one for the reference category ("cauc" in the example above). This is typically what is required for fitting econometric regression models, and hence changing the contrasts is usually not necessary.

The function I()

Some further details on the specification of regression models via Wilkinson-Rogers type formulas in R are in order. We have already seen that the arithmetic operator + has a different meaning in formulas: it is employed to add regressors (main effects). Additionally, the operators :, *, /, ^ have special meanings, all related to the specification of interaction effects (see the following section).

To be able to use the arithmetic operators in their original meaning in a formula, they can be protected from the formula interpretation by insulating them inside a function, as in `log(x1 * x2)`. If the problem at hand does not require a transformation, R's `I()` function can be used, which returns its argument "as is". This was used for computing experience squared in the regression above.

Comparison of models

With more than a single explanatory variable, it is interesting to test for the relevance of subsets of regressors. For any two nested models, this can be done using the function `anova()`, which we already encountered in the preceding section. For example, it might be desired to test for the relevance of the variable `ethnicity`; i.e., whether there is a difference in the average log-wage (controlling for experience and education) between Caucasian and African-American men. As we have used treatment contrasts for `ethnicity`, the significance of this effect can already be read off the summary shown above. However, to illustrate the general procedure for model comparisons, we explicitly fit the model without `ethnicity` and then compare both models using `anova()`:

```
R> cps_noeth <- lm(log(wage) ~ experience + I(experience^2) +
+    education, data = CPS1988)
R> anova(cps_noeth, cps_lm)

Analysis of Variance Table

Model 1: log(wage) ~ experience + I(experience^2) + education
Model 2: log(wage) ~ experience + I(experience^2) +
  education + ethnicity
  Res.Df  RSS    Df Sum of Sq   F Pr(>F)
1  28151 9720
2  28150 9599     1      121  355 <2e-16
```

This reveals that the effect of `ethnicity` is significant at any reasonable level.

The usage of `anova()` shown above is slightly different from the usage we illustrated in the previous section. If several fitted models are supplied to `anova()`, the associated RSS are compared (in the order in which the models are entered). If only a single model is passed to `anova()`, as in the preceding section, then it sequentially adds the terms in the order specified by the formula (starting from the trivial model with only an intercept) and compares the associated RSS:

```
R> anova(cps_lm)

Analysis of Variance Table
```

```
Response: log(wage)
                  Df Sum Sq Mean Sq F value Pr(>F)
experience         1    840     840    2462 <2e-16
I(experience^2)    1   2249    2249    6597 <2e-16
education          1   1620    1620    4750 <2e-16
ethnicity          1    121     121     355 <2e-16
Residuals      28150   9599    0.34
```

The next to last line in this ANOVA table is equivalent to the direct comparison of cps_lm and cps_noeth.

There is also a more elegant way to fit the latter model given the former. It is not necessary to type in the model formula again. Instead, the update() method for "lm" objects can be used. It takes as its arguments a fitted "lm" object and the changes in the model formula relative to the original specification. Hence

```
R> cps_noeth <- update(cps_lm, formula = . ~ . - ethnicity)
```

also yields the same fitted-model object cps_noeth, which can then be used in anova() as shown above. The expression . ~ . - ethnicity specifies to take the LHS and RHS in the formula (signaled by the "."), only removing ethnicity on the RHS.

Finally, the updating via update() and the model comparison via anova() can be done in one step if the function waldtest() from the package lmtest (Zeileis and Hothorn 2002) is used. The package is loaded automatically by **AER**, and waldtest() can be used as in

```
R> waldtest(cps_lm, . ~ . - ethnicity)
```

```
Wald test
```

```
Model 1: log(wage) ~ experience + I(experience^2) +
  education + ethnicity
Model 2: log(wage) ~ experience + I(experience^2) + education
  Res.Df    Df   F Pr(>F)
1  28150
2  28151    -1 355 <2e-16
```

By default, this produces the same F test as anova() but does not report the associated RSS. The reason is that waldtest() can also perform quasi-F tests in situations where errors are potentially heteroskedastic. This is described in more detail in Chapter 4.

3.3 Partially Linear Models

Quadratic terms in experience are common in wage equations; however, given the size of the CPS1988 data, it may be worthwhile to model the role of this variable using more flexible tools. In R, this is particularly easy, and so we briefly consider a semiparametric extension, the partially linear model

$$\texttt{log(wage)} = \beta_1 + g(\texttt{experience}) + \beta_2\,\texttt{education} + \beta_3\,\texttt{ethnicity} + \varepsilon$$

Here, g is an unknown function to be estimated from the data, and we use regression splines for this task (see, e.g., Hastie, Tibshirani, and Friedman 2001). In R, splines are available in the package **splines** (part of the base R distribution and automatically loaded with **AER**). Among the many available types, B splines are computationally convenient, and the function providing them is bs(). It can directly be used in lm(), and so fitting the desired model is as easy as

```
R> cps_plm <- lm(log(wage) ~ bs(experience, df = 5) +
+     education + ethnicity, data = CPS1988)
```

We omit the summary, as the coefficients on the spline basis are not easily interpreted, and only note that the return on education for this specification is 8.82% per year. There are a couple of possibilities for using bs(): one can either specify the **degree** of the piecewise polynomial (defaulting to 3) and the **knots** by hand or supply the parameter **df**, which selects the remaining ones. The expression bs(experience, df = 5) thus internally generates regressors, namely piecewise cubic polynomials evaluated at the observations pertaining to experience, with the implied $5-3 = 2$ interior knots evenly spaced and therefore located at the 33.33% and 66.67% quantiles of experience.

The choice of df = 5 was made based on model selection in terms of the Schwarz criterion (BIC). The following code considers B splines with df ranging from 3 through 10, suggesting that df = 5 is appropriate for the data at hand:

```
R> cps_bs <- lapply(3:10, function(i) lm(log(wage) ~
+     bs(experience, df = i) + education + ethnicity,
+     data = CPS1988))
R> structure(sapply(cps_bs, AIC, k = log(nrow(CPS1988))),
+     .Names = 3:10)
```

3	4	5	6	7	8	9	10
49205	48836	48794	48795	48801	48797	48799	48802

First, a list cps_bs of fitted linear models is constructed via lapply(), to which extractor functions can be easily applied as in sapply(cps_bs, AIC). The call above utilizing structure() is slightly more complicated because it additionally sets $\log(n)$ as the penalty parameter (BIC instead of the default AIC) and assigns intuitive names (degrees of freedom) to the resulting vector.

The cubic spline from the selected model is best compared with the classical fit from cps_lm by means of a graphical display. The following code plots log-wage as a function of experience (for Caucasian workers with average years of education) for the classical model with a quadratic term in experience and for the partially linear model.

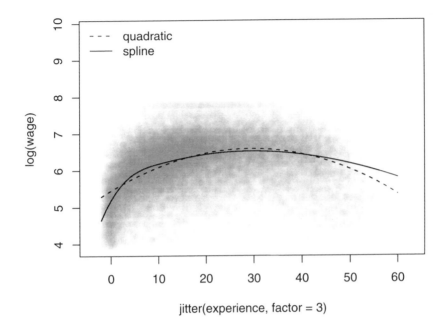

Fig. 3.4. Partially linear model versus classical specification for the CPS1988 data.

```
R> cps <- data.frame(experience = -2:60, education =
+    with(CPS1988, mean(education[ethnicity == "cauc"])),
+    ethnicity = "cauc")
R> cps$yhat1 <- predict(cps_lm, newdata = cps)
R> cps$yhat2 <- predict(cps_plm, newdata = cps)
R> plot(log(wage) ~ jitter(experience, factor = 3), pch = 19,
+    col = rgb(0.5, 0.5, 0.5, alpha = 0.02), data = CPS1988)
R> lines(yhat1 ~ experience, data = cps, lty = 2)
R> lines(yhat2 ~ experience, data = cps)
R> legend("topleft", c("quadratic", "spline"), lty = c(2,1),
+    bty = "n")
```

Figure 3.4 indicates that both models are not too distinct for the 20–40 years of experience range. Overall, the spline version exhibits less curvature beyond eight years of experience. However, the most remarkable feature of the plot appears to be the more pronounced curvature below seven years of experience in the spline fit, which is largely missed by the traditional approach. An alternative approach to partially linear models is to use kernel methods, this is implemented in the package **np** (Hayfield and Racine 2008).

Some further remarks on the plot are appropriate. The large number of observations and numerous ties in `experience` provide a challenge in that the standard scatterplot will result in overplotting. Here, we circumvent the problem by (1) adding some amount of "jitter" (i.e., noise) to the regressor `experience` and (2) setting the color to "semitransparent" gray. This results in darker shades of gray for areas with more data points, thus conveying a sense of density. This technique is also called "alpha blending" and requires that, in addition to the color itself, a value of `alpha`—ranging between 0 (fully transparent) and 1 (opaque)—be specified. Various color functions in R provide an argument `alpha`; e.g., the basic `rgb()` function implementing the RGB (red, green, blue) color model. Selecting equal intensities for all three color values in `rgb()` yields a shade of gray (which would be more conveniently available in `gray()`, but this does not allow for alpha blending).

Note that alpha transparency is not available for all plotting devices in R. Among others, it is available for `windows()` (typically used on Microsoft Windows), `quartz()` (typically used on Mac OS X), and `pdf()`, provided that the argument `version` is set to `version = "1.4"` or greater (on all platforms). See `?rgb` for further details. A somewhat simpler but less appealing solution available on all devices is to employ the default color (i.e., black) and a tiny plotting character such as `pch = "."`.

3.4 Factors, Interactions, and Weights

In labor economics, there exist many empirical studies trying to shed light on the issue of discrimination (for example, by gender or ethnicity). These works typically involve regressions with factors and interactions. Since the `CPS1988` data contain the factor `ethnicity`, we consider some more general specifications of the basic wage equation (3.2) in order to see whether there are aspects of parameter heterogeneity that need to be taken into account. Of course, our example is merely an illustration of working with factors and interactions, and we do not seriously address any discrimination issues.

Technically, we are interested in the empirical relevance of an interaction between `ethnicity` and other variables in our regression model. Before doing so for the data at hand, the most important specifications of interactions in R are outlined in advance.

The operator `:` specifies an interaction effect that is, in the default contrast coding, essentially the product of a dummy variable and a further variable (possibly also a dummy). The operator `*` does the same but also includes the corresponding main effects. The same is done by `/`, but it uses a nested coding instead of the interaction coding. Finally, `^` can be used to include all interactions up to a certain order within a group of variables. Table 3.2

Table 3.2. Specification of interactions in formulas.

Formula	Description
y ~ a + x	Model without interaction: identical slopes with respect to x but different intercepts with respect to a.
y ~ a * x y ~ a + x + a:x	Model with interaction: the term a:x gives the difference in slopes compared with the reference category.
y ~ a / x y ~ a + x %in% a	Model with interaction: produces the same fitted values as the model above but using a nested coefficient coding. An explicit slope estimate is computed for each category in a.
y ~ (a + b + c)^2 y ~ a*b*c - a:b:c	Model with all two-way interactions (excluding the three-way interaction).

provides a brief overview for numerical variables y, x and categorical variables a, b, c.

Interactions

The factor `ethnicity` is already included in our model; however, at present it only affects the intercept. A priori, it is not clear whether slope coefficients are also affected; i.e., whether Caucasians and African-Americans are paid differently conditional on some further regressors. For illustration, let us consider an interaction between `ethnicity` and `education`. In R, this is conveniently specified as in the following call:

```
R> cps_int <- lm(log(wage) ~ experience + I(experience^2) +
+     education * ethnicity, data = CPS1988)
R> coeftest(cps_int)
```

```
t test of coefficients:
```

	Estimate	Std. Error	t value	Pr(>|t|)
(Intercept)	4.313059	0.019590	220.17	<2e-16
experience	0.077520	0.000880	88.06	<2e-16
I(experience^2)	-0.001318	0.000019	-69.34	<2e-16
education	0.086312	0.001309	65.94	<2e-16
ethnicityafam	-0.123887	0.059026	-2.10	0.036
education:ethnicityafam	-0.009648	0.004651	-2.07	0.038

We see that the interaction term is statistically significant at the 5% level. However, with a sample comprising almost 30,000 individuals, this can hardly be taken as compelling evidence for inclusion of the term.

Above, just the table of coefficients and associated tests is computed for compactness. This can be done using coeftest() (instead of summary()); see Chapter 4 for further details.

As described in Table 3.2, the term education*ethnicity specifies inclusion of three terms: ethnicity, education, and the interaction between the two (internally, the product of the dummy indicating ethnicity=="afam" and education). Specifically, education*ethnicity may be thought of as expanding to 1 + education + ethnicity + education:ethnicity; the coefficients are, in this order, the intercept for Caucasians, the slope for education for Caucasians, the difference in intercepts, and the difference in slopes. Hence, the interaction term is also available without inclusion of ethnicity and education, namely as education:ethnicity. Thus, the following call is equivalent to the preceding one, though somewhat more clumsy:

```
R> cps_int <- lm(log(wage) ~ experience + I(experience^2) +
+     education + ethnicity + education:ethnicity,
+     data = CPS1988)
```

Separate regressions for each level

As a further variation, it may be necessary to fit separate regressions for African-Americans and Caucasians. This could either be done by computing two separate "lm" objects using the subset argument to lm() (e.g., lm(formula, data, subset = ethnicity=="afam", ...) or, more conveniently, by using a single linear-model object in the form

```
R> cps_sep <- lm(log(wage) ~ ethnicity /
+     (experience + I(experience^2) + education) - 1,
+     data = CPS1988)
```

This model specifies that the terms within parentheses are nested within ethnicity. Here, an intercept is not needed since it is best replaced by two separate intercepts for the two levels of ethnicity; the term -1 removes it. (Note, however, that the R^2 is computed differently in the summary(); see ?summary.lm for details.)

For compactness, we just give the estimated coefficients for the two groups defined by the levels of ethnicity:

```
R> cps_sep_cf <- matrix(coef(cps_sep), nrow = 2)
R> rownames(cps_sep_cf) <- levels(CPS1988$ethnicity)
R> colnames(cps_sep_cf) <- names(coef(cps_lm))[1:4]
R> cps_sep_cf
```

```
      (Intercept) experience I(experience^2) education
cauc        4.310    0.07923      -0.0013597    0.08575
afam        4.159    0.06190      -0.0009415    0.08654
```

This shows that the effects of education are similar for both groups, but the remaining coefficients are somewhat smaller in absolute size for African-Americans.

A comparison of the new model with the first one yields

```
R> anova(cps_sep, cps_lm)
```

Analysis of Variance Table

```
Model 1: log(wage) ~ ethnicity/(experience +
  I(experience^2) + education) - 1
Model 2: log(wage) ~ experience + I(experience^2) +
  education + ethnicity
  Res.Df  RSS    Df Sum of Sq    F  Pr(>F)
1  28147 9582
2  28150 9599    -3      -17 16.5 1.1e-10
```

Hence, the model where ethnicity interacts with every other regressor fits significantly better, at any reasonable level, than the model without any interaction term. (But again, this is a rather large sample.)

Change of the reference category

In any regression containing an (unordered) factor, R by default uses the first level of the factor as the reference category (whose coefficient is fixed at zero). In CPS1988, "cauc" is the reference category for ethnicity, while "northeast" is the reference category for region.

Bierens and Ginther (2001) employ "south" as the reference category for region. For comparison with their article, we now change the contrast coding of this factor, so that "south" becomes the reference category. This can be achieved in various ways; e.g., by using contrasts() or by simply changing the order of the levels in the factor. As the former offers far more complexity than is needed here (but is required, for example, in statistical experimental design), we only present a solution using the latter. We set the reference category for region in the CPS1988 data frame using relevel() and subsequently fit a model in which this is included:

```
R> CPS1988$region <- relevel(CPS1988$region, ref = "south")
R> cps_region <- lm(log(wage) ~ ethnicity + education +
+     experience + I(experience^2) + region, data = CPS1988)
R> coef(cps_region)
```

```
     (Intercept)   ethnicityafam       education      experience
        4.283606       -0.225679        0.084672        0.077656
 I(experience^2)  regionnortheast   regionmidwest      regionwest
       -0.001323        0.131920        0.043789        0.040327
```

Weighted least squares

Cross-section regressions are often plagued by heteroskedasticity. Diagnostic tests against this alternative will be postponed to Chapter 4, but here we illustrate one of the remedies, weighted least squares (WLS), in an application to the journals data considered in Section 3.1. The reason is that `lm()` can also handle weights.

Readers will have noted that the upper left plot of Figure 3.3 already indicates that heteroskedasticity is a problem with these data. A possible solution is to specify a model of conditional heteroskedasticity, e.g.

$$E(\varepsilon_i^2 | x_i, z_i) = g(z_i^\top \gamma),$$

where g, the skedastic function, is a nonlinear function that can take on only positive values, z_i is an ℓ-vector of observations on exogenous or predetermined variables, and γ is an ℓ-vector of parameters.

Here, we illustrate the fitting of some popular specifications using the price per citation as the variable z_i. Recall that assuming $E(\varepsilon_i^2 | x_i, z_i) = \sigma^2 z_i^2$ leads to a regression of y_i/z_i on $1/z_i$ and x_i/z_i. This means that the fitting criterion changes from $\sum_{i=1}^n (y_i - \beta_1 - \beta_2 x_i)^2$ to $\sum_{i=1}^n z_i^{-2}(y_i - \beta_1 - \beta_2 x_i)^2$, i.e., each term is now weighted by z_i^{-2}. The solutions $\hat{\beta}_1, \hat{\beta}_2$ of the new minimization problem are called the weighted least squares (WLS) estimates, a special case of generalized least squares (GLS). In R, this model is fitted using

```
R> jour_wls1 <- lm(log(subs) ~ log(citeprice), data = journals,
+     weights = 1/citeprice^2)
```

Note that the weights are entered as they appear in the new fitting criterion. Similarly,

```
R> jour_wls2 <- lm(log(subs) ~ log(citeprice), data = journals,
+     weights = 1/citeprice)
```

yields a regression with weights of the form $1/z_i$. Figure 3.5 provides the OLS regression line along with the lines corresponding to the two WLS specifications:

```
R> plot(log(subs) ~ log(citeprice), data = journals)
R> abline(jour_lm)
R> abline(jour_wls1, lwd = 2, lty = 2)
R> abline(jour_wls2, lwd = 2, lty = 3)
R> legend("bottomleft", c("OLS", "WLS1", "WLS2"),
+     lty = 1:3, lwd = 2, bty = "n")
```

More often than not, we are not sure as to which form of the skedastic function to use and would prefer to estimate it from the data. This leads to feasible generalized least squares (FGLS).

In our case, the starting point could be

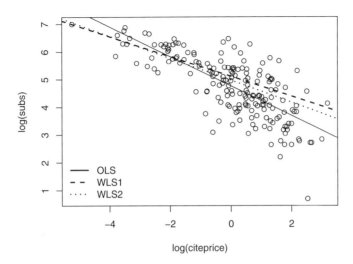

Fig. 3.5. Scatterplot of the journals data with least-squares (solid) and weighted least-squares (dashed and dotted) lines.

$$E(\varepsilon_i^2|x_i) = \sigma^2 x_i^{\gamma_2} = \exp(\gamma_1 + \gamma_2 \log x_i),$$

which we estimate by regressing the logarithm of the squared residuals from the OLS regression on the logarithm of `citeprice` and a constant. In the second step, we use the fitted values of this auxiliary regression as the weights in the model of interest:

```
R> auxreg <- lm(log(residuals(jour_lm)^2) ~ log(citeprice),
+     data = journals)
R> jour_fgls1 <- lm(log(subs) ~ log(citeprice),
+     weights = 1/exp(fitted(auxreg)), data = journals)
```

It is possible to iterate further, yielding a second variant of the FGLS approach. A compact solution makes use of a while loop:

```
R> gamma2i <- coef(auxreg)[2]
R> gamma2 <- 0
R> while(abs((gamma2i - gamma2)/gamma2) > 1e-7) {
+     gamma2 <- gamma2i
+     fglsi <- lm(log(subs) ~ log(citeprice), data = journals,
+       weights = 1/citeprice^gamma2)
+     gamma2i <- coef(lm(log(residuals(fglsi)^2) ~
+       log(citeprice), data = journals))[2]
```

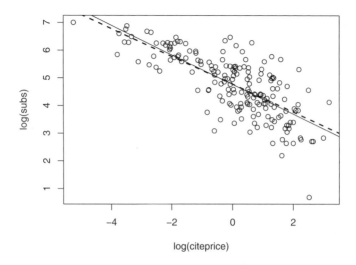

Fig. 3.6. Scatterplot of the journals data with OLS (solid) and iterated FGLS (dashed) lines.

```
+  }
R> jour_fgls2 <- lm(log(subs) ~ log(citeprice), data = journals,
+    weights = 1/citeprice^gamma2)
```

This loop specifies that, as long as the relative change of the slope coefficient γ_2 exceeds 10^{-7} in absolute value, the iteration is continued; that is, a new set of coefficients for the skedastic function is computed utilizing the residuals from a WLS regression that employs the skedastic function estimated in the preceding step. The final estimate of the skedastic function resulting from the while loop is then used in a further WLS regression, whose coefficients are referred to as iterated FGLS estimates. This approach yields

```
R> coef(jour_fgls2)
```

```
(Intercept) log(citeprice)
     4.7758        -0.5008
```

and the parameter **gamma2** of the transformation equals 0.2538, quite distinct from our first attempts using a predetermined skedastic function.

Figure 3.6 provides the OLS regression line along with the line corresponding to the iterated FGLS estimator. We see that the iterated FGLS solution is more similar to the OLS solution than to the various WLS specifications considered before.

3.5 Linear Regression with Time Series Data

In econometrics, time series regressions are often fitted by OLS. Hence, in principle, they can be fitted like any other linear regression model using `lm()` if the data set is held in a "`data.frame`". However, this is typically not the case for time series data, which are more conveniently stored in one of R's time series classes. An example is "`ts`", which holds its data in a vector or matrix plus some time series attributes (start, end, frequency). More detailed information on time series analysis in R is provided in Chapter 6. Here, it suffices to note that using `lm()` with "`ts`" series has two drawbacks: (1) for fitted values or residuals, the time series properties are by default not preserved, and (2) lags or differences cannot directly be specified in the model formula.

These problems can be tackled in different ways. The simplest solution is to do the additional computations "by hand"; i.e., to compute lags or differences before calling `lm()`. Alternatively, the package **dynlm** (Zeileis 2008) provides the function `dynlm()`, which tries to overcome the problems described above.[1] It allows for formulas such as `d(y) ~ L(d(y)) + L(x, 4)`, here describing a regression of the first differences of a variable y on its first difference lagged by one period and on the fourth lag of a variable x; i.e., $y_i - y_{i-1} = \beta_1 + \beta_2(y_{i-1} - y_{i-2}) + \beta_3 x_{i-4} + \varepsilon_i$. This is an autoregressive distributed lag (ADL) model.

As an illustration, we will follow Greene (2003, Chapter 8) and consider different forms for a consumption function based on quarterly US macroeconomic data from 1950(1) through 2000(4) as provided in the data set `USMacroG`, a "`ts`" time series. For such objects, there exists a `plot()` method, here employed for visualizing disposable income `dpi` and `consumption` (in billion USD) via

```
R> data("USMacroG")
R> plot(USMacroG[, c("dpi", "consumption")], lty = c(3, 1),
+     plot.type = "single", ylab = "")
R> legend("topleft", legend = c("income", "consumption"),
+     lty = c(3, 1), bty = "n")
```

The result is shown in Figure 3.7. Greene (2003) considers two models,

$$\text{consumption}_i = \beta_1 + \beta_2 \, \text{dpi}_i + \beta_3 \, \text{dpi}_{i-1} + \varepsilon_i$$
$$\text{consumption}_i = \beta_1 + \beta_2 \, \text{dpi}_i + \beta_3 \, \text{consumption}_{i-1} + \varepsilon_i.$$

In the former model, a distributed lag model, consumption responds to changes in income only over two periods, while in the latter specification, an autoregressive distributed lag model, the effects of income changes persist due to the autoregressive specification. The models can be fitted to the `USMacroG` data by `dynlm()` as follows:

---

[1] A different approach that also works for modeling functions other than `lm()` is implemented in the package **dyn** (Grothendieck 2005).

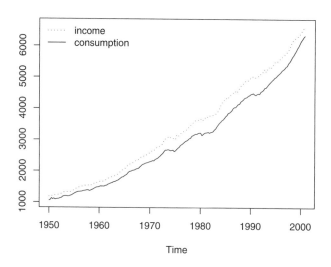

Fig. 3.7. Time series plot of the US consumption and income series (in billion USD).

```
R> library("dynlm")
R> cons_lm1 <- dynlm(consumption ~ dpi + L(dpi), data = USMacroG)
R> cons_lm2 <- dynlm(consumption ~ dpi + L(consumption),
+     data = USMacroG)
```

The corresponding summaries are of the same type as for "lm" objects. In addition, the sampling period used is reported at the beginning of the output:

```
R> summary(cons_lm1)

Time series regression with "ts" data:
Start = 1950(2), End = 2000(4)

Call:
dynlm(formula = consumption ~ dpi + L(dpi),
  data = USMacroG)

Residuals:
    Min      1Q  Median      3Q     Max
-190.02  -56.68    1.58   49.91  323.94

Coefficients:
            Estimate Std. Error t value Pr(>|t|)
```

```
(Intercept) -81.0796     14.5081    -5.59  7.4e-08
dpi           0.8912      0.2063     4.32  2.4e-05
L(dpi)        0.0309      0.2075     0.15   0.88
```

```
Residual standard error: 87.6 on 200 degrees of freedom
Multiple R-squared: 0.996,        Adjusted R-squared: 0.996
F-statistic: 2.79e+04 on 2 and 200 DF,  p-value: <2e-16
```

The second model fits the data slightly better. Here only lagged consumption but not income is significant:

```
R> summary(cons_lm2)
```

```
Time series regression with "ts" data:
Start = 1950(2), End = 2000(4)
```

```
Call:
dynlm(formula = consumption ~ dpi + L(consumption),
  data = USMacroG)
```

```
Residuals:
    Min      1Q  Median     3Q     Max
 -101.30   -9.67   1.14   12.69   45.32
```

```
Coefficients:
                Estimate Std. Error t value Pr(>|t|)
(Intercept)      0.53522    3.84517    0.14     0.89
dpi             -0.00406    0.01663   -0.24     0.81
L(consumption)   1.01311    0.01816   55.79   <2e-16
```

```
Residual standard error: 21.5 on 200 degrees of freedom
Multiple R-squared:     1,        Adjusted R-squared:     1
F-statistic: 4.63e+05 on 2 and 200 DF,  p-value: <2e-16
```

The RSS of the first model can be obtained with deviance(cons_lm1) and equals 1534001.49, and the RSS of the second model, computed by deviance(cons_lm2), equals 92644.15. To visualize these two fitted models, we employ time series plots of fitted values and residuals (see Figure 3.8) showing that the series of the residuals of cons_lm1 is somewhat U-shaped, while that of cons_lm2 is closer to zero. To produce this plot, the following code was used:

```
R> plot(merge(as.zoo(USMacroG[,"consumption"]), fitted(cons_lm1),
+     fitted(cons_lm2), 0, residuals(cons_lm1),
+     residuals(cons_lm2)), screens = rep(1:2, c(3, 3)),
+     lty = rep(1:3, 2), ylab = c("Fitted values", "Residuals"),
+     xlab = "Time", main = "")
```

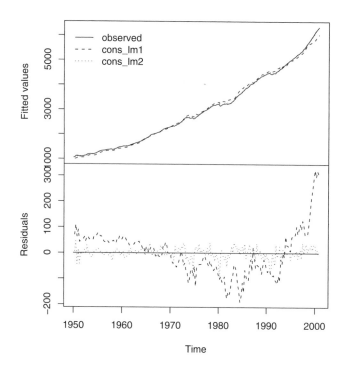

Fig. 3.8. Fitted models for US consumption.

```
R> legend(0.05, 0.95, c("observed", "cons_lm1", "cons_lm2"),
+     lty = 1:3, bty = "n")
```

This is somewhat complicated at first sight, but the components are rather easy to understand: first, we merge() the original series with the fitted values from both models, a zero line and the residuals of both models. The merged series is plotted subsequently on two screens with different line types and some more annotation. Before merging, the original series (of class "ts") is coerced to class "zoo" from the package **zoo** via as.zoo(), a generalization of "ts" with a slightly more flexible plot() method. The class "zoo" is discussed in more detail in Chapter 6.

Encompassing test

To discriminate between these two competing nonnested models, we consider an encompassing test. Alternative methods to perform this task are also available; we consider the J test and the Cox test (see Davidson and MacKinnon 1981, 2004) in an exercise. All of these tests implement slightly different approaches to testing the same null hypotheses.

The encompassing approach to comparing two nonnested models is to transform the problem into a situation we can already deal with: comparing nested models. The idea is to fit the encompassing model comprising all regressors from both competing models, in our case the autoregressive distributed lag (ADL) model

```
R> cons_lmE <- dynlm(consumption ~ dpi + L(dpi) +
+    L(consumption), data = USMacroG)
```

and then to compare each of the two nonnested models with the encompassing model. Now if one of the models is not significantly worse than the encompassing model while the other is, this test would favor the former model over the latter. As illustrated in the previous sections, nested models can be compared with anova(), and we can even carry out both tests of interest in one go:

```
R> anova(cons_lm1, cons_lmE, cons_lm2)
```

```
Analysis of Variance Table

Model 1: consumption ~ dpi + L(dpi)
Model 2: consumption ~ dpi + L(dpi) + L(consumption)
Model 3: consumption ~ dpi + L(consumption)
  Res.Df     RSS  Df Sum of Sq       F  Pr(>F)
1    200 1534001
2    199   73550   1   1460451  3951.4 < 2e-16
3    200   92644  -1    -19094    51.7 1.3e-11
```

The first F test compares the model cons_lm1 with the encompassing model cons_lmE, and the second F test compares cons_lmE and cons_lm2. Both models perform significantly worse compared with the encompassing model, although the F statistic is much smaller for cons_lm2.

Instead of computing the encompassing model by hand and then calling anova(), the function encomptest() from the **lmtest** package can be used. This simplifies the call to

```
R> encomptest(cons_lm1, cons_lm2)
```

```
Encompassing test

Model 1: consumption ~ dpi + L(dpi)
Model 2: consumption ~ dpi + L(consumption)
Model E: consumption ~ dpi + L(dpi) + L(consumption)
          Res.Df  Df       F  Pr(>F)
M1 vs. ME    199  -1  3951.4 < 2e-16
M2 vs. ME    199  -1    51.7 1.3e-11
```

and leads to equivalent output. Internally, `encomptest()` relies on the `waldtest()` function mentioned above.

3.6 Linear Regression with Panel Data

There has been considerable interest in panel data econometrics over the last two decades, and hence it is almost mandatory to include a brief discussion of some common specifications in R. The package **plm** (Croissant and Millo 2008) contains the relevant fitting functions and methods. For the methodological background, we refer to Baltagi (2005).

Static linear models

For illustrating the basic fixed- and random-effects methods, we use the well-known Grunfeld data (Grunfeld 1958) comprising 20 annual observations on the three variables real gross investment (`invest`), real value of the firm (`value`), and real value of the capital stock (`capital`) for 11 large US firms for the years 1935–1954. Originally employed in a study of the determinants of corporate investment in a University of Chicago Ph.D. thesis, these data have been a textbook classic since the 1970s. The accompanying package **AER** provides the full data set comprising all 11 firms, and the documentation contains further details on alternative versions and errors therein.

The main difference between cross-sectional data and panel data is that panel data have an internal structure, indexed by a two-dimensional array, which must be communicated to the fitting function. We refer to the cross-sectional objects as "individuals" and the time identifier as "time".

We use a subset of three firms for illustration and, utilizing `plm.data()`, tell R that the individuals are called `"firm"`, whereas the time identifier is called `"year"`:

```
R> data("Grunfeld", package = "AER")
R> library("plm")
R> gr <- subset(Grunfeld, firm %in% c("General Electric",
+     "General Motors", "IBM"))
R> pgr <- plm.data(gr, index = c("firm", "year"))
```

Instead of setting up a structured data frame `pgr` in advance, this could also be specified on the fly in calls to `plm()` simply by supplying `index = c("firm", "year")`. For later use, plain OLS on the pooled data is available using

```
R> gr_pool <- plm(invest ~ value + capital, data = pgr,
+     model = "pooling")
```

The basic one-way panel regression is

$$\texttt{invest}_{it} = \beta_1 \texttt{value}_{it} + \beta_2 \texttt{capital}_{it} + \alpha_i + \nu_{it}, \qquad (3.3)$$

where $i = 1, \ldots, n$, $t = 1, \ldots, T$, and the α_i denote the individual-specific effects.

A fixed-effects version is estimated by running OLS on a within-transformed model:

```
R> gr_fe <- plm(invest ~ value + capital, data = pgr,
+    model = "within")
R> summary(gr_fe)

Oneway (individual) effect Within Model

Call:
plm(formula = invest ~ value + capital, data = pgr,
  model = "within")

Balanced Panel: n=3, T=20, N=60

Residuals :
  Min. 1st Qu.  Median 3rd Qu.    Max.
-167.00  -26.10    2.09   26.80  202.00

Coefficients :
        Estimate Std. Error t-value Pr(>|t|)
value    0.1049    0.0163    6.42  1.3e-10
capital  0.3453    0.0244   14.16  < 2e-16

Total Sum of Squares: 1890000
Residual Sum of Squares: 244000
Multiple R-Squared: 0.871
F-statistic: 185.407 on 55 and 2 DF, p-value: 0.00538
```

The summary provides a tabulation of observations per firm, here indicating that the data are balanced, as well as the usual regression output with coefficients (excluding the fixed effects) and associated standard errors and some measures of goodness of fit. A two-way model could have been estimated upon setting effect = "twoways". If fixed effects need to be inspected, a fixef() method and an associated summary() method are available.

It is of interest to check whether the fixed effects are really needed. This is done by comparing the fixed effects and the pooled OLS fits by means of pFtest() and yields

```
R> pFtest(gr_fe, gr_pool)

    F test for effects

data:  invest ~ value + capital
F = 56.82, df1 = 2, df2 = 55, p-value = 4.148e-14
alternative hypothesis: significant effects
```

indicating that there is substantial interfirm variation.

It is also possible to fit a random-effects version of (3.3) using the same fitting function upon setting `model = "random"` and selecting a method for estimating the variance components. Recall that a random-effects estimator is essentially an FGLS estimator, utilizing OLS after "quasi-demeaning" all variables, where the precise form of the quasi-demeaning transformation depends on the `random.method` selected. Four methods are available: Swamy-Arora, Amemiya, Wallace-Hussain, and Nerlove (see, e.g., Baltagi 2005, for further information on these estimators). The default is `random.method = "swar"` (for Swamy-Arora), and for illustration we use `random.method = "walhus"` (for Wallace-Hussain) and obtain

```
R> gr_re <- plm(invest ~ value + capital, data = pgr,
+     model = "random", random.method = "walhus")
R> summary(gr_re)

Oneway (individual) effect Random Effect Model
(Wallace-Hussain's transformation)

Call:
plm(formula = invest ~ value + capital, data = pgr,
  model = "random", random.method = "walhus")

Balanced Panel: n=3, T=20, N=60

Effects:
                var std.dev share
idiosyncratic 4389.3    66.3  0.35
individual    8079.7    89.9  0.65
theta:   0.837

Residuals :
   Min. 1st Qu.  Median 3rd Qu.    Max.
-187.00  -32.90    6.96   31.40  210.00

Coefficients :
              Estimate Std. Error t-value Pr(>|t|)
(intercept) -109.9766    61.7014   -1.78    0.075
value          0.1043     0.0150    6.95 3.6e-12
capital        0.3448     0.0245   14.06  < 2e-16

Total Sum of Squares: 1990000
Residual Sum of Squares: 258000
Multiple R-Squared: 0.87
F-statistic: 191.545 on 57 and 2 DF, p-value: 0.00521
```

A comparison of the regression coefficients shows that fixed- and random-effects methods yield rather similar results for these data.

As was the case with the fixed-effects regression, it is of interest to check whether the random effects are really needed. Several versions of a Lagrange multiplier test for this task are available in `plmtest()`, defaulting to the test proposed by Honda (1985). Here it yields

```
R> plmtest(gr_pool)

        Lagrange Multiplier Test - (Honda)

data:  invest ~ value + capital
normal = 15.47, p-value < 2.2e-16
alternative hypothesis: significant effects
```

This test also suggests that some form of parameter heterogeneity must be taken into account.

Random-effects methods are more efficient than the fixed-effects estimator under more restrictive assumptions, namely exogeneity of the individual effects. It is therefore important to test for endogeneity, and the standard approach employs a Hausman test. The relevant function `phtest()` requires two panel regression objects, in our case yielding

```
R> phtest(gr_re, gr_fe)

        Hausman Test

data:  invest ~ value + capital
chisq = 0.0404, df = 2, p-value = 0.98
alternative hypothesis: one model is inconsistent
```

In line with the rather similar estimates presented above, endogeneity does not appear to be a problem here.

plm contains methods for further types of static panel data models, notably the Hausman-Taylor model (Hausman and Taylor 1981) and varying coefficient models. We consider the Hausman-Taylor estimator in an exercise.

Dynamic linear models

To conclude this section, we present a more advanced example, the dynamic panel data model

$$y_{it} = \sum_{j=1}^{p} \varrho_j y_{i,t-j} + x_{it}^{\top}\beta + u_{it}, \quad u_{it} = \alpha_i + \beta_t + \nu_{it}, \qquad (3.4)$$

estimated by the method of Arellano and Bond (1991). Recall that their estimator is a generalized method of moments (GMM) estimator utilizing lagged endogenous regressors after a first-differences transformation. **plm** comes with the original Arellano-Bond data (EmplUK) dealing with determinants of employment (emp) in a panel of 140 UK firms for the years 1976–1984. The data are unbalanced, with seven to nine observations per firm.

To simplify the exposition, we first set up a static formula containing the relevant variables average annual wage per employee (wage), the book value of gross fixed assets (capital), and an index of value-added output at constant factor cost (output), all in logarithms:

```
R> data("EmplUK", package = "plm")
R> form <- log(emp) ~ log(wage) + log(capital) + log(output)
```

The function providing the Arellano-Bond estimator is pgmm(), and it takes as its first argument a so-called dynformula, this being a static model equation, as given above, augmented with a list of integers containing the number of lags for each variable. The dynamic employment equation from Arellano and Bond (1991) is now estimated as follows:

```
R> empl_ab <- pgmm(dynformula(form, list(2, 1, 0, 1)),
+      data = EmplUK, index = c("firm", "year"),
+      effect = "twoways", model = "twosteps",
+      gmm.inst = ~ log(emp), lag.gmm = list(c(2, 99)))
```

Hence, this is a dynamic model with $p = 2$ lagged endogenous terms (compare this with Equation (3.4)) in which both log(wage) and log(output) occur up to lag 1, while for log(capital) only the contemporaneous term is included. A model containing time as well as firm-specific effects is specified, the instruments are lagged terms of the dependent variable log(emp), and the lags used are given by the argument lag.gmm = list(c(2, 99)), indicating that all lags beyond lag 1 are to be used as instruments. This results in

```
R> summary(empl_ab)

Twoways effects Two steps model

Call:
pgmm(formula = log(emp) ~ lag(log(emp), 1) + lag(log(emp),
   2) + log(wage) + lag(log(wage), 1) + log(capital) +
   log(output) + lag(log(output), 1), data = EmplUK,
   effect = "twoways", model = "twosteps",
   gmm.inst = ~log(emp), lag.gmm = list(c(2, 99)),
   index = c("firm", "year"))

Unbalanced Panel: n=140, T=7-9, N=1031

Number of Observations Used:  611
```

```
Residuals
     Min.    1st Qu.    Median      Mean    3rd Qu.      Max.
-0.619000 -0.049500 -0.000457 -0.000184  0.053400  0.641000
```

```
Coefficients
                     Estimate Std. Error z-value Pr(>|z|)
lag(log(emp), 1)       0.4742     0.0853    5.56  2.7e-08
lag(log(emp), 2)      -0.0530     0.0273   -1.94  0.05222
log(wage)             -0.5132     0.0493  -10.40  < 2e-16
lag(log(wage), 1)      0.2246     0.0801    2.81  0.00502
log(capital)           0.2927     0.0395    7.42  1.2e-13
log(output)            0.6098     0.1085    5.62  1.9e-08
lag(log(output), 1)   -0.4464     0.1248   -3.58  0.00035
```

```
Sargan Test: chisq(25) = 30.11 (p.value=0.22)
Autocorrelation test (1): normal = -2.428 (p.value=0.0076)
Autocorrelation test (2): normal = -0.3325 (p.value=0.37)
Wald test for coefficients: chisq(7) = 372 (p.value=<2e-16)
Wald test for time dummies: chisq(6) = 26.90 (p.value=0.000151)
```

suggesting that autoregressive dynamics are important for these data. The tests at the bottom of the output indicate that the model is not fully satisfactory, but we refrain from presenting Arellano and Bond's preferred specification, which in addition treats wages and capital as endogenous.

Note that, due to constructing lags and taking first differences, three cross sections are lost; hence the estimation period is 1979–1984 and only 611 observations are effectively available for estimation.

3.7 Systems of Linear Equations

Systems of regression equations have been a hallmark of econometrics for several decades. Standard examples include seemingly unrelated regressions and various macroeconomic simultaneous equation models. The package **systemfit** (Henningsen and Hamann 2007) can estimate a number of multiple-equation models. As an example, we present a seemingly unrelated regression (SUR) model (Zellner 1962) for the Grunfeld data. As noted by Greene (2003, p. 329, fn. 39), "[a]lthough admittedly not current, these data are unusually cooperative for illustrating the different aspects of estimating systems of regression equations". Unlike the panel data models considered in the preceding section, which permit only individual-specific intercepts, the SUR model also allows for individual-specific slopes. (As regards terminology, the "individuals" of the preceding section will now be referred to as "equations".) The model assumes contemporaneous correlation across equations, and thus joint estimation of all parameters is, in general, more efficient than OLS on each equation.

The main fitting function is `systemfit()`. When fitting a SUR model, it requires a "`plm.data`" object containing the information on the internal structure of the data. This is set up as in the preceding section utilizing `plm.data()` (to save space we only consider two firms):

```
R> library("systemfit")
R> gr2 <- subset(Grunfeld, firm %in% c("Chrysler", "IBM"))
R> pgr2 <- plm.data(gr2, c("firm", "year"))
```

The main arguments to `systemfit()` are a formula, a data set, and a `method` (defaulting to `"OLS"`). Here we need `method = "SUR"`:

```
R> gr_sur <- systemfit(invest ~ value + capital,
+     method = "SUR", data = pgr2)
R> summary(gr_sur, residCov = FALSE, equations = FALSE)

systemfit results
method: SUR

          N DF  SSR detRCov OLS-R2 McElroy-R2
system 40 34 4114   11022  0.929      0.927

          N DF  SSR   MSE  RMSE    R2 Adj R2
Chrysler 20 17 3002 176.6 13.29 0.913  0.903
IBM      20 17 1112  65.4  8.09 0.952  0.946


Coefficients:
                     Estimate Std. Error t value Pr(>|t|)
Chrysler_(Intercept)  -5.7031    13.2774   -0.43  0.67293
Chrysler_value         0.0780     0.0196    3.98  0.00096
Chrysler_capital       0.3115     0.0287   10.85  4.6e-09
IBM_(Intercept)       -8.0908     4.5216   -1.79  0.09139
IBM_value              0.1272     0.0306    4.16  0.00066
IBM_capital            0.0966     0.0983    0.98  0.33951
```

Thus, `summary()` provides the standard regression results for each equation in a compact layout as well as some measures of overall fit. For compactness, we suppressed parts of the rather voluminous default output. Readers may want to run this example with `summary(gr_sur)` to obtain more detailed information, including between-equation correlations.

The output indicates again that there is substantial variation among the firms, and thus a single-equation model for the pooled data is not appropriate.

In addition to SUR models, **systemfit** can estimate linear simultaneous-equations models by several methods (two-stage least squares, three-stage least squares, and variants thereof), and there is also a fitting function for certain nonlinear specifications.

3.8 Exercises

1. This exercise is taken from Faraway (2005, p. 23). Generate some artificial data by
   ```
   x <- 1:20
   y <- x + rnorm(20)
   ```
   Fit a polynomial in x for predicting y. Compute $\hat{\beta}$ in two ways—by `lm()` and by using the direct calculation $\hat{\beta} = (X^\top X)^{-1} X^\top y$. At what degree of polynomial does the direct calculation method fail? (Note the need for the `I()` function in fitting the polynomial; e.g., in `lm(y ~ x + I(x^2))`.) (Lesson: *Never* use the textbook formula $\hat{\beta} = (X^\top X)^{-1} X^\top y$ for computations!)

2. Estimate a hedonic regression for the `HousePrices` data taken from Anglin and Gençay (1996), which contain prices of houses sold in the city of Windsor, Canada, during July, August, and September 1987. These data are also used in the textbook by Verbeek (2004).
   (a) Fit a multiple linear regression model to the logarithm of the price, using all remaining variables as regressors. Experiment with models containing lot size, number of bathrooms, number of bedrooms, and stories in logarithms and in levels, respectively. Which model do you prefer?
   (b) What is the expected price of a two-story house of 4,700 sq. ft. with three bedrooms, two bathrooms, a driveway, no recreational room, a full finished basement, without gas heating or air conditioning, and two-car garage, that is not located in a preferred area? Report also a prediction interval.
   (c) In order to determine whether the logarithmic transformation of the dependent variable is really appropriate, a Box-Cox transformation might be helpful. Use the function `boxcox()` from the package **MASS**. What do you conclude?

3. Consider the `PSID1982` data from Cornwell and Rupert (1988) and discussed further in Baltagi (2002).
   (a) Regress the logarithm of the wage on all available regressors plus experience squared.
   (b) Does gender interact with education and/or experience?

4. Section 3.5 considered two competing models for US consumption utilizing an encompassing test. Different approaches to comparing nonnested models are the J test suggested by Davidson and MacKinnon (1981) and the Cox test. Both are available in the package **lmtest** in the functions `jtest()` and `coxtest()`. For the methodological background, we refer to Greene (2003, Chapter 8) and Davidson and MacKinnon (2004).
   (a) Test `cons_lm1` vs. `cons_lm2` using the J test.
   (b) Test `cons_lm1` vs. `cons_lm2` using the Cox test.
   Do all tests lead to similar conclusions?

5. Use the PSID1982 data and consider the following two nonnested models (compare with Baltagi 2002, p. 230):

$$M_1 : \texttt{log(wage)} = \beta_0 + \beta_1\texttt{education} + \beta_2\texttt{experience} + \beta_3\texttt{experience}^2$$
$$+\beta_4\texttt{weeks} + \beta_5\texttt{married} + \beta_6\texttt{gender}$$
$$+\beta_7\texttt{ethnicity} + \beta_8\texttt{union} + \varepsilon$$
$$M_2 : \texttt{log(wage)} = \beta_0 + \beta_1\texttt{education} + \beta_2\texttt{experience} + \beta_3\texttt{experience}^2$$
$$+\beta_4\texttt{weeks} + \beta_5\texttt{occupation} + \beta_6\texttt{south}$$
$$+\beta_7\texttt{smsa} + \beta_8\texttt{industry} + \nu$$

(a) Compute the J tests for M_1 vs. M_2 and M_2 vs. M_1, respectively.
(b) Both M_1 and M_2 can be artificially nested within a larger model. Use an F test for M_1 versus this augmented model. Repeat for M_2 versus the augmented model. What do you conclude?

6. The estimator of Hausman and Taylor (1981) is appropriate when only some of the individual effects in a panel regression are endogenous. Employ this estimator on a wage equation for the PSID1982 data using all 12 regressors appearing in the preceding exercise. Note that you will have to enter all exogenous variables as instruments for themselves.

(a) Consider the regressors experience, experience^2, occupation, industry, union, and education as endogenous (as do Cornwell and Rupert 1988).
(b) Consider the regressors experience, experience^2, weeks, married, union, and education as endogenous (as do Baltagi and Khanti-Akom 1990; Baltagi 2005, Table 7.4).
(c) Which estimates do you consider more plausible?

7. The function gls() from the package **nlme** will fit one of the classical econometric regression models, the linear regression model $y_i = x_i^\top \beta + \varepsilon_i$ with AR(1) disturbances, $\varepsilon_i = \phi\varepsilon_{i-1} + \nu_i$, where $\nu_i \sim (0, \sigma_\nu^2)$ i.i.d. and $|\phi| < 1$, albeit by maximum likelihood, not by least-squares techniques. Select one of the firms from the Grunfeld data and estimate this model. What is the amount of first-order autocorrelation?

8. Find a way of estimating the SUR model from Section 3.7 using the **plm** package.

9. The function ivreg() from package **AER** will fit instrumental variable (IV) regressions. Using the USConsump1993 data taken from Baltagi (2002), estimate the simple Keynesian consumption function

$$\texttt{expenditure}_t = \beta_0 + \beta_1\texttt{income}_t + \varepsilon_t$$

(a) by OLS.
(b) by IV. The only available instrument is investment, given by the identity expenditure + investment = income.
(c) Compare both sets of estimates using a Hausman test, thus replicating Baltagi (2002, Section 11.7). What do you conclude?

4

Diagnostics and Alternative Methods of Regression

The techniques presented in the preceding chapter tell only part of the story of regression analysis. On the one hand, all of the estimates, tests, and other summaries are computed as if the model and its assumptions were correct, but on the other, there exist various regression methods apart from OLS, some of which are more appropriate in certain applications.

Here, we discuss several approaches to validating linear regression models:

- A popular approach (on occasion somewhat misleading) compares various statistics computed for the full data set with those obtained from deleting single observations. This is known as regression diagnostics.
- In econometrics, diagnostic tests have played a prominent role since about 1980. The most important alternative hypotheses are heteroskedasticity, autocorrelation, and misspecification of the functional form.
- Also, the impenetrable disturbance structures typically present in observational data have led to the development of "robust" covariance matrix estimators, a number of which have become available during the last 20 years.

This chapter provides a brief introduction to all three approaches. It turns out that graphical techniques are often quite effective at revealing structure that one may not have suspected.

Furthermore, resistant (in the sense of resistance to outliers and unusual observations) regression techniques are often quite helpful, although such methods do not appear to be as widely known among econometricians as they deserve to be. We also include a brief introduction to quantile regression, a method that has been receiving increasing interest in applied econometrics, especially in labor economics.

C. Kleiber, A. Zeileis, *Applied Econometrics with R*,
DOI: 10.1007/978-0-387-77318-6_4, © Springer Science+Business Media, LLC 2008

4.1 Regression Diagnostics

There is extensive literature on examining the fit of linear regression models, mostly under the label "regression diagnostics". The goal is to find points that are not fitted as well as they should be or have undue influence on the fitting of the model. The techniques associated with Belsley, Kuh, and Welsch (1980) based on deletion of observations are widely available in statistical software packages, and R is no exception.

This topic provides an excellent opportunity to further illustrate some basic aspects of R programming, showing that variations on available functions or plots are easily constructed with just a few lines of code. For illustration, we consider the PublicSchools data from the **sandwich** package, taken from Greene (1993). They provide, for the 50 states of the United States of America and for Washington, DC, per capita Expenditure on public schools and per capita Income by state, all for the year 1979:

```
R> data("PublicSchools")
R> summary(PublicSchools)
```

```
   Expenditure         Income
 Min.    :259    Min.    : 5736
 1st Qu.:315    1st Qu.: 6670
 Median :354    Median : 7597
 Mean    :373    Mean    : 7608
 3rd Qu.:426    3rd Qu.: 8286
 Max.    :821    Max.    :10851
 NA's    :  1
```

We first omit the incomplete observations using na.omit()—this affects only Wisconsin, where Expenditure is not available. Subsequently, we generate a scatterplot with a fitted linear model, in which three observations are highlighted:

```
R> ps <- na.omit(PublicSchools)
R> ps$Income <- ps$Income / 10000
R> plot(Expenditure ~ Income, data = ps, ylim = c(230, 830))
R> ps_lm <- lm(Expenditure ~ Income, data = ps)
R> abline(ps_lm)
R> id <- c(2, 24, 48)
R> text(ps[id, 2:1], rownames(ps)[id], pos = 1, xpd = TRUE)
```

The resulting plot in Figure 4.1 shows that there is a positive relationship between expenditure and income; however, Alaska seems to be a dominant observation in that it is far away from the bulk of the data. Two further observations, Washington, DC, and Mississippi, also appear to deserve a closer look. Visual inspection suggests that the OLS regression line is somewhat tilted. We will pursue this issue below.

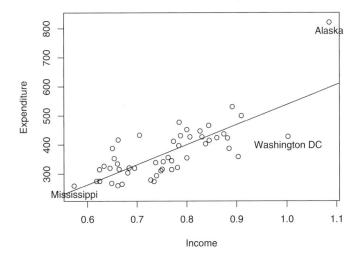

Fig. 4.1. Per capita expenditure on public schools versus per capita income.

The basic tool for examining the fit is the residuals, and we have already seen that `plot()` applied to an "lm" object yields several diagnostic plots. Figure 4.2, resulting from

```
R> plot(ps_lm, which = 1:6)
```

depicts all six types (currently) available. The default is `which = c(1:3,5)`; i.e., only four are shown.

Specifically, the first plot depicts residuals versus fitted values. It is useful for checking the assumption $E(\varepsilon|X) = 0$, and it brings out systematic variations in the residuals that suggest that the model is incomplete. The tests and confidence intervals used in `summary(ps_lm)` are based on the assumption of i.i.d. normal errors. The residuals can be assessed for normality using a QQ plot comparing the residuals to "ideal" normal observations. It plots the ordered residuals against $\Phi^{-1}(i/(n+1))$, $i = 1, \ldots, n$, where Φ^{-1} is the quantile function of the standard normal distribution. The third plot, a scale-location plot, depicts $\sqrt{|\hat{r}_i|}$ (for the standardized residuals r_i; see below for details) against $\hat{y}_i$. It is useful for checking the assumption that errors are identically distributed, in particular that the variance is homogeneous, $\mathsf{Var}(\varepsilon|X) = \sigma^2 I$. The remaining three plots display several combinations of standardized residuals, leverage, and Cook's distance—these diagnostic measures will be discussed in more detail below.

Thus, Figure 4.2 demonstrates that Alaska stands out in all plots: it has a large residual (top left), and it therefore appears in the upper tail of the

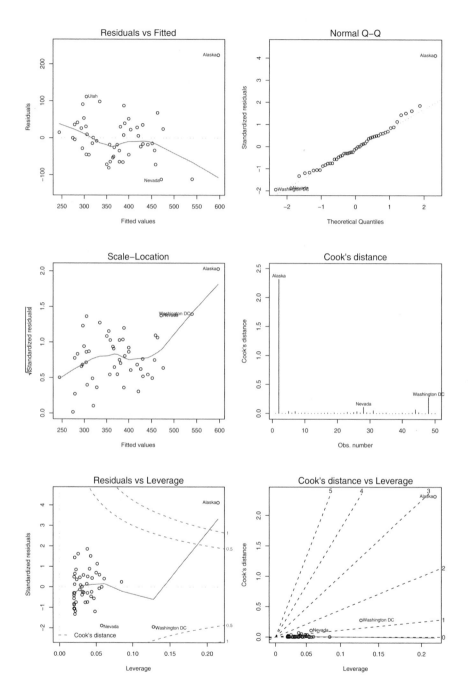

Fig. 4.2. Diagnostic plots for the public schools regression.

empirical distribution of the residuals (top right), it casts doubt on the assumption of homogeneous variances (middle left), it corresponds to an extraordinarily large Cook's distance (middle and bottom right), and it has the highest leverage (bottom left and right). There are further observations singled out, but none of these are as dominant as Alaska. Clearly, these observations deserve a closer look. To further explore the fit of our model, we briefly review regression diagnostics for linear models.

Leverage and standardized residuals

Recall that least-squares residuals are not independent and that they do not have the same variance. Specifically, if $\mathsf{Var}(\varepsilon|X) = \sigma^2 I$, their variance-covariance matrix is $\mathsf{Var}(\hat{\varepsilon}|X) = \sigma^2(I - H)$, where $H = X(X^\top X)^{-1} X^\top$ is the "hat matrix".

Basic diagnostics are the diagonal elements h_{ii} of H, which in R are provided by the generic function `hatvalues()`. Since $\mathsf{Var}(\hat{\varepsilon}_i|X) = \sigma^2(1 - h_{ii})$, observations with large values of h_{ii} will have small values of $\mathsf{Var}(\hat{\varepsilon}_i|X)$, and hence residuals for such observations tend to be close to zero. Therefore, h_{ii} is said to measure the *leverage* of the observation i. The trace of H is k (the number of regressors), and "large" is commonly taken to mean greater than two or three times the average element on the diagonal of H, k/n. Note that the value of h_{ii} depends only on X and not on y, and hence leverages contain only partial information about a point. This implies that there are two kinds of high-leverage points: "bad leverage points" and "good leverage points". The former correspond to observations with large h_{ii} and unusual y_i; such observations can have a dramatic impact on the fit of a model. The latter correspond to observations with large h_{ii} and typical y_i; such observations are beneficial because they improve the precision of the estimates.

Figure 4.3, obtained via

```
R> ps_hat <- hatvalues(ps_lm)
R> plot(ps_hat)
R> abline(h = c(1, 3) * mean(ps_hat), col = 2)
R> id <- which(ps_hat > 3 * mean(ps_hat))
R> text(id, ps_hat[id], rownames(ps)[id], pos = 1, xpd = TRUE)
```

depicts the diagonal elements of the hat matrix for the `PublicSchools` data. $\mathsf{Var}(\hat{\varepsilon}_i|X) = \sigma^2(1 - h_{ii})$ also suggests the use of

$$ r_i = \frac{\hat{\varepsilon}_i}{\hat{\sigma}\sqrt{1 - h_{ii}}}. $$

The r_i are the standardized residuals, available via `rstandard()` in R. (They are referred to as internally studentized residuals by some authors. This should not be confused with the (externally) studentized residuals defined below.) If the model assumptions are correct, $\mathsf{Var}(r_i|X) = 1$ and $\mathsf{Cor}(r_i, r_j|X)$ tends to be small.

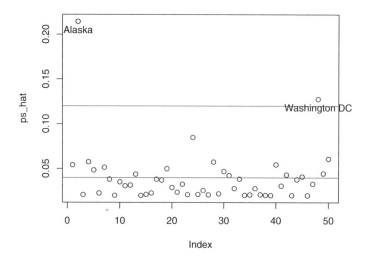

Fig. 4.3. Diagonal elements of the hat matrix for the public schools data, with their mean and three times their mean as horizontal lines.

Deletion diagnostics

To detect unusual observations, Belsley *et al.* (1980) propose to compute various leave-one-out (or deletion) diagnostics. Thus, one excludes point i and computes, for example, the estimates $\hat{\beta}_{(i)}$ and $\hat{\sigma}_{(i)}$, where the subscript (i) denotes that observation i has been excluded. Similarly, $\hat{y}_{(i)} = X\hat{\beta}_{(i)}$ denotes the vector of predictions of the y_i utilizing $\hat{\beta}_{(i)}$. Any observation whose removal from the data would cause a large change in the fit is considered "influential". It may or may not have large leverage and may or may not be an outlier, but it will tend to have at least one of these properties.

Following Belsley *et al.* (1980), basic quantities are

$$DFFIT_i = y_i - \hat{y}_{i,(i)},$$

$$DFBETA = \hat{\beta} - \hat{\beta}_{(i)},$$

$$COVRATIO_i = \frac{\det(\hat{\sigma}^2_{(i)}(X^\top_{(i)}X_{(i)})^{-1})}{\det(\hat{\sigma}^2(X^\top X)^{-1})},$$

$$D^2_i = \frac{(\hat{y} - \hat{y}_{(i)})^\top(\hat{y} - \hat{y}_{(i)})}{k\hat{\sigma}^2}.$$

Here, *DFFIT* measures the change in the fitted values, while *DFBETA* measures the changes in the coefficients. *COVRATIO* considers the change in the estimate of the OLS covariance matrix, while the D_i^2, the Cook's distances (already encountered in the diagnostic plots in Figure 4.2), are especially popular because they reduce the information to a single value for each observation. Appropriately scaled versions of the first two quantities are called *DFFITS* and *DFBETAS*.

All these objects are available in the corresponding R functions dffit(), dffits(), dfbeta(), dfbetas(), covratio(), and cooks.distance(). In addition, the function rstudent() provides the (externally) studentized residuals

$$t_i = \frac{\hat{\varepsilon}_i}{\hat{\sigma}_{(i)}\sqrt{1 - h_{ii}}}$$

alluded to above.

The function influence.measures()

Fortunately, it is not necessary to compute all the preceding quantities separately (although this is possible). R provides the convenience function influence.measures(), which simultaneously calls dfbetas(), dffits(), covratio(), and cooks.distance(), returning a rectangular array of these diagnostics. In addition, it highlights observations that are unusual for at least one of the influence measures. Since the common influence measures are functions of r_i and/or h_{ii}, which ones to choose is often a matter of taste. Since

```
R> influence.measures(ps_lm)
```

provides a rectangular array of size 50×6, we do not show the result. Instead we briefly consider selected aspects.

With two regressors and 50 observations, the average diagonal element of the hat matrix is 0.04. The points with leverage higher than three times the mean leverage can be obtained via

```
R> which(ps_hat > 3 * mean(ps_hat))

    Alaska Washington DC
      2            48
```

This highlights Alaska and Washington, DC; for Alaska, the leverage is even larger than five times the mean (see Figure 4.3).

If the observations that are (potentially) influential according to at least one criterion are desired, use

```
R> summary(influence.measures(ps_lm))

Potentially influential observations of
       lm(formula = Expenditure ~ Income, data = ps) :
```

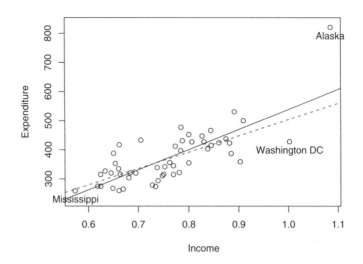

Fig. 4.4. Least-squares line with (solid) and without (dashed) Alaska, Mississippi, and Washington, DC.

	dfb.1_	dfb.Incm	dffit	cov.r	cook.d	hat
Alaska	-2.39_*	2.52_*	2.65_*	0.55_*	2.31_*	0.21_*
Mississippi	0.07	-0.07	0.08	1.14_*	0.00	0.08
Washington DC	0.66	-0.71	-0.77_*	1.01	0.28	0.13_*

This again points us to Alaska and Washington, DC, and in addition Mississippi, which had been highlighted in Figure 4.1. It is noteworthy that Alaska stands out by any measure of influence. From Figure 4.4, resulting from

```
R> plot(Expenditure ~ Income, data = ps, ylim = c(230, 830))
R> abline(ps_lm)
R> id <- which(apply(influence.measures(ps_lm)$is.inf, 1, any))
R> text(ps[id, 2:1], rownames(ps)[id], pos = 1, xpd = TRUE)
R> ps_noinf <- lm(Expenditure ~ Income, data = ps[-id,])
R> abline(ps_noinf, lty = 2)
```

it is clear that this observation is a bad leverage point. To some extent, this is also true for Washington, DC, but here the situation is not nearly as bad as in the case of Alaska. The plot also depicts the least-squares line excluding the three influential points, and it is obvious that it provides a much better summary of the data.

4.2 Diagnostic Tests

A more formal approach to validating regression models is diagnostic testing. Many cross-section regressions are plagued by heteroskedasticity. Similarly, regression models for time series data should be tested for residual autocorrelation. The package **lmtest** (Zeileis and Hothorn 2002), originally inspired by the methods discussed in Krämer and Sonnberger (1986), provides a number of tests for both alternatives plus further tests for misspecification of the functional form. A discussion of the underlying theory is provided in Baltagi (2002), Davidson and MacKinnon (2004), and Greene (2003), to mention a few sources. We proceed to the practical aspects of these procedures and refer the reader to these sources for the background formalities.

Almost all of the tests discussed below return an object of class "htest" (standing for <u>h</u>ypothesis <u>test</u>), containing the value of the test statistic, the corresponding p value, additional parameters such as degrees of freedom (where appropriate), the name of the tested model, and perhaps the method used if there are several variants of the test.

For cross-section regressions, we return to the Journals data used in the preceding chapter. As before, we set up a reduced data set `journals` that also contains the `age` of the journals (for the year 2000, when the data were collected).

```
R> data("Journals")
R> journals <- Journals[, c("subs", "price")]
R> journals$citeprice <- Journals$price/Journals$citations
R> journals$age <- 2000 - Journals$foundingyear
```

As before, we consider a simple model that explains the logarithm of the number of library subscriptions for a journal by the logarithm of the price per citation.

```
R> jour_lm <- lm(log(subs) ~ log(citeprice), data = journals)
```

For a scatterplot of the data and the fitted model, we refer to Figure 3.1.

Testing for heteroskedasticity

For cross-section regressions, the assumption $\mathsf{Var}(\varepsilon_i|x_i) = \sigma^2$ is typically in doubt. A popular test for checking this assumption is the Breusch-Pagan test (Breusch and Pagan 1979). It fits a linear regression model to the squared residuals $\hat{\varepsilon}_i^2$ of the model under investigation and rejects if too much of the variance is explained by the additional explanatory variables. For this auxiliary regression, the same explanatory variables X are taken as in the main model. Other strategies could be to use the fitted values $\hat{y}_i$ or the original regressors plus squared terms and interactions (White 1980). For our model fitted to the journals data, stored in `jour_lm`, the diagnostic plots in Figure 3.3 suggest that the variance decreases with the fitted values or, equivalently, it increases with the price per citation. Hence, the regressor log(citeprice) used in the main model should also be employed for the auxiliary regression.

Under H_0, the test statistic of the Breusch-Pagan test approximately follows a χ_q^2 distribution, where q is the number of regressors in the auxiliary regression (excluding the constant term). The textbook version of the test is for normally distributed errors, an assumption that is often too strong for practical work. It can be overcome by using a studentized version (Koenker 1981).

The function bptest() implements all these flavors of the Breusch-Pagan test. By default, it computes the studentized statistic for the auxiliary regression utilizing the original regressors X. Hence

```
R> bptest(jour_lm)

        studentized Breusch-Pagan test

data:  jour_lm
BP = 9.803, df = 1, p-value = 0.001742
```

uses log(citeprice) in the auxiliary regression and detects heteroskedasticity in the data with respect to the price per citation. Alternatively, the White test picks up the heteroskedasticity. It uses the original regressors as well as their squares and interactions in the auxiliary regression, which can be passed as a second formula to bptest(). Here, this reduces to

```
R> bptest(jour_lm, ~ log(citeprice) + I(log(citeprice)^2),
+     data = journals)

        studentized Breusch-Pagan test

data:  jour_lm
BP = 10.91, df = 2, p-value = 0.004271
```

Another test for heteroskedasticity—nowadays probably more popular in textbooks than in applied work—is the Goldfeld-Quandt test (Goldfeld and Quandt 1965). Its simple idea is that after ordering the sample with respect to the variable explaining the heteroskedasticity (e.g., price per citation in our example), the variance at the beginning should be different from the variance at the end of the sample. Hence, splitting the sample and comparing the mean residual sum of squares before and after the split point via an F test should be able to detect a change in the variance. However, there are few applications where a meaningful split point is known in advance; hence, in the absence of better alternatives, the center of the sample is often used. Occasionally, some central observations are omitted in order to improve the power. The function gqtest() implements this test; by default, it assumes that the data are already ordered, and it also uses the middle of the sample as the split point without omitting any central observations.

In order to apply this test to the journals data, we order the observations with respect to price per citation. This leads again to a highly significant result that confirms the Breusch-Pagan analysis presented above:

```
R> gqtest(jour_lm, order.by = ~ citeprice, data = journals)

        Goldfeld-Quandt test

data:  jour_lm
GQ = 1.703, df1 = 88, df2 = 88, p-value = 0.00665
```

Testing the functional form

The assumption $E(\varepsilon|X) = 0$ is crucial for consistency of the least-squares estimator. A typical source for violation of this assumption is a misspecification of the functional form; e.g., by omitting relevant variables. One strategy for testing the functional form is to construct auxiliary variables and assess their significance using a simple F test. This is what Ramsey's RESET (regression specification error test; Ramsey 1969) does: it takes powers of the fitted values $\hat{y}$ and tests whether they have a significant influence when added to the regression model. Alternatively, powers of the original regressors or of the first principal component of X can be used. All three versions are implemented in the function resettest(). It defaults to using second and third powers of the fitted values as auxiliary variables. With only one real regressor in the model matrix X (excluding the intercept), all three strategies yield equivalent results. Hence we use

```
R> resettest(jour_lm)

        RESET test

data:  jour_lm
RESET = 1.441, df1 = 2, df2 = 176, p-value = 0.2395
```

to assess whether second and third powers of log(citeprice) can significantly improve the model. The result is clearly non-significant, and hence no misspecification can be detected in this direction.

The rainbow test (Utts 1982) takes a different approach to testing the functional form. Its basic idea is that even a misspecified model might be able to fit the data reasonably well in the "center" of the sample but might lack fit in the tails. Hence, the rainbow test fits a model to a subsample (typically the middle 50%) and compares it with the model fitted to the full sample using an F test. To determine the "middle", the sample has to be ordered; e.g., by a regressor variable or by the Mahalanobis distance of the regressor vector x_i to the mean regressor. Both procedures are implemented (along with some further options for fine-tuning the choice of the subsample) in the function raintest(). For the jour_lm model, we may use

```
R> raintest(jour_lm, order.by = ~ age, data = journals)

        Rainbow test
```

```
data:  jour_lm
Rain = 1.774, df1 = 90, df2 = 88, p-value = 0.003741
```

This orders the journals data by age and assesses whether the model fit for the 50% "middle-aged" journals is different from the fit comprising all journals. This appears to be the case, signaling that the relationship between the number of subscriptions and the price per citation also depends on the age of the journal. As we will see below, the reason seems to be that libraries are willing to pay more for established journals.

Another diagnostic test that relies on ordering the sample prior to testing is the Harvey-Collier test (Harvey and Collier 1977). For the ordered sample, the test computes the recursive residuals (Brown, Durbin, and Evans 1975) of the fitted model. The recursive residuals are essentially standardized one-step-ahead prediction errors: the linear model is fitted to the first $i-1$ observations, and from this the ith observation is predicted and the corresponding residual is computed and suitably standardized. If the model is correctly specified, the recursive residuals have mean zero, whereas the mean should significantly differ from zero if the ordering variable has an influence on the regression relationship. Therefore, the Harvey-Collier test is a simple t test for zero mean; it is implemented in the function harvtest(). Applying this test to the journals data yields the following result:

```
R> harvtest(jour_lm, order.by = ~ age, data = journals)

        Harvey-Collier test
```

```
data:  jour_lm
HC = 5.081, df = 177, p-value = 9.464e-07
```

This confirms the results from the rainbow test, emphasizing that the age of the journals has a significant influence on the regression relationship.

Testing for autocorrelation

Just as the disturbances in cross-section models are typically heteroskedastic, they are often affected by autocorrelation (or serial correlation) in time series regressions. Let us reconsider the first model for the US consumption function from Section 3.5:

```
R> data("USMacroG")
R> consump1 <- dynlm(consumption ~ dpi + L(dpi),
+    data = USMacroG)
```

A classical testing procedure suggested for assessing autocorrelation in regression relationships is the Durbin-Watson test (Durbin and Watson 1950). The test statistic is computed as the ratio of the sum of the squared first differences of the residuals (i.e., $(\hat{\varepsilon}_i - \hat{\varepsilon}_{i-1})^2$), and the residual sum of squares. Under the null hypothesis of no autocorrelation, the test statistic should be in the vicinity of 2—under the alternative of positive autocorrelation, it typically is much smaller. The distribution under the null hypothesis is nonstandard: under the assumption of normally distributed errors, it can be represented as the distribution of a linear combination of χ^2 variables with weights depending on the regressor matrix X. Many textbooks still recommend using tabulated upper and lower bounds of critical values. The function dwtest() implements an exact procedure for computing the p value and also provides a normal approximation for sufficiently large samples (both depending on the regressor matrix X). The function can be easily applied to the fitted consumption function via

```
R> dwtest(consump1)

        Durbin-Watson test

data:  consump1
DW = 0.0866, p-value < 2.2e-16
alternative hypothesis: true autocorrelation is greater than 0
```

here detecting a highly significant positive autocorrelation, which confirms the results from Section 3.5.

Further tests for autocorrelation, originally suggested for diagnostic checking of ARIMA models (see Chapter 6), are the Box-Pierce test (Box and Pierce 1970) and a modified version, the Ljung-Box test (Ljung and Box 1978). Both are implemented in the function Box.test() in base R (in the **stats** package). The test statistics are (approximate) χ^2 statistics based on estimates of the autocorrelations up to order p: the Box-Pierce statistic is n times the sum of squared autocorrelations, and the Ljung-Box refinement weighs the squared autocorrelation at lag j by $(n + 2)/(n - j)$ $(j = 1, \ldots, p)$. The default in Box.test() is to use the Box-Pierce version with $p = 1$. Unlike the diagnostic tests in **lmtest**, the function expects a series of residuals and not the specification of a linear model as its primary argument. Here, we apply the Box-Ljung test to the residuals of consump1, which again yields a highly significant result.

```
R> Box.test(residuals(consump1), type = "Ljung-Box")

        Box-Ljung test

data:  residuals(consump1)
X-squared = 176.1, df = 1, p-value < 2.2e-16
```

An alternative approach to assessing autocorrelation is the Breusch-Godfrey test (Breusch 1979; Godfrey 1978)—unlike the Durbin-Watson test, this also works in the presence of lagged dependent variables (see also Section 7.1). The Breusch-Godfrey test is an LM test against both AR(p) and

MA(p) alternatives, computed by fitting an auxiliary regression that explains the residuals $\hat{\varepsilon}$ by the original regressors X augmented by the lagged residuals up to order p ($\hat{\varepsilon}_{i-1}, \ldots, \hat{\varepsilon}_{i-p}$) (where zeros are used as starting values). The resulting RSS is compared with the RSS of the original RSS in a χ^2 (or F) test. Both versions are implemented in the function bgtest(). By default, it uses an order of $p = 1$. Here, we obtain

```
R> bgtest(consump1)
```

```
        Breusch-Godfrey test for serial correlation of order 1

data:  consump1
LM test = 193.0, df = 1, p-value < 2.2e-16
```

again confirming the previous results for this model.

More details on various time series models in R and dealing with autocorrelation are provided in Chapter 6.

4.3 Robust Standard Errors and Tests

As illustrated in the preceding sections, economic data typically exhibit some form of autocorrelation and/or heteroskedasticity. If the covariance structure were known, it could be taken into account in a (parametric) model, but more often than not, the form of the autocorrelation or heteroskedasticity is unknown. In such cases, regression coefficients can typically still be estimated consistently using OLS (given that the functional form is correctly specified), but for valid inference a consistent covariance matrix estimate is essential. Over the last 20 years, several procedures for heteroskedasticity consistent (HC) and, more generally, for heteroskedasticity and autocorrelation consistent (HAC) covariance matrix estimation have been suggested in the econometrics literature. These are today routinely applied in econometric practice.

To be more specific, the problem is that the standard t and F tests performed when calling summary() or anova() for a fitted linear model assume that errors are homoskedastic and uncorrelated given the regressors: $\mathsf{Var}(\varepsilon|X) = \sigma^2 I$. In practice, however, as motivated for the data sets above, this is often not the case and $\mathsf{Var}(\varepsilon|X) = \Omega$, where Ω is unknown. In this situation, the covariance matrix of the estimated regression coefficients $\hat{\beta}$ is given by

$$\mathsf{Var}(\hat{\beta}|X) \quad = \quad \left(X^\top X\right)^{-1} X^\top \Omega X \left(X^\top X\right)^{-1},$$

which only reduces to the familiar $\sigma^2 (X^\top X)^{-1}$ if the errors are indeed homoskedastic and uncorrelated. Hence, for valid inference in models with heteroskedasticity and/or autocorrelation, it is vital to compute a robust estimate of $\mathsf{Var}(\hat{\beta}|X)$. In R, the package **sandwich** (which is automatically attached when loading **AER**) provides the functions vcovHC() and vcovHAC()

implementing HC and HAC counterparts of vcov(). Furthermore, the estimates produced by these functions can be easily plugged into the functions coeftest() and waldtest() from **lmtest**, functions generalizing the summary() and anova() methods for fitted linear models. More details on the implementation and application of these functions can be found in Zeileis (2004).

HC estimators

In cross-section regressions, it is typically appropriate to assume that Ω is a diagonal matrix with potentially non-constant diagonal entries. Therefore, a natural plug-in estimator for $\mathsf{Var}(\hat{\beta}|X)$ could use $\hat{\Omega} = \mathrm{diag}(\omega_1, \ldots, \omega_n)$; that is, a diagonal matrix. Various choices for ω_i have been suggested in the literature:

$$\mathrm{const}: \quad \omega_i = \hat{\sigma}^2$$

$$\mathrm{HC0}: \quad \omega_i = \hat{\varepsilon}_i^2$$

$$\mathrm{HC1}: \quad \omega_i = \frac{n}{n-k}\,\hat{\varepsilon}_i^2$$

$$\mathrm{HC2}: \quad \omega_i = \frac{\hat{\varepsilon}_i^2}{1 - h_{ii}}$$

$$\mathrm{HC3}: \quad \omega_i = \frac{\hat{\varepsilon}_i^2}{(1 - h_{ii})^2}$$

$$\mathrm{HC4}: \quad \omega_i = \frac{\hat{\varepsilon}_i^2}{(1 - h_{ii})^{\delta_i}}$$

where h_{ii} are again the diagonal elements of the hat matrix, $\bar{h}$ is their mean, and $\delta_i = \min\{4, h_{ii}/\bar{h}\}$. All these variants are available in the function vcovHC().

The first version is the standard estimator for homoskedastic errors. All others produce different kinds of HC estimators. The estimator HC0 was introduced by Eicker (1963) and popularized in econometrics by White (1980). The estimators HC1, HC2, and HC3 were suggested by MacKinnon and White (1985) to improve the performance in small samples. A more extensive study of small-sample behavior was conducted by Long and Ervin (2000), who arrive at the conclusion that HC3 provides the best performance in small samples, as it gives less weight to influential observations. HC3 is the default choice in vcovHC(). More recently, Cribari-Neto (2004) suggested the estimator HC4 to further improve small-sample performance, especially in the presence of influential observations.

The function vcovHC() can be applied to a fitted linear model, just like the function vcov(). For the model fitted to the journals data, both yield rather similar estimates of the covariance matrix:

```
R> vcov(jour_lm)
```

```
              (Intercept) log(citeprice)
(Intercept)      3.126e-03      -6.144e-05
log(citeprice)  -6.144e-05       1.268e-03
```

R> vcovHC(jour_lm)

```
              (Intercept) log(citeprice)
(Intercept)      0.003085       0.000693
log(citeprice)   0.000693       0.001188
```

To obtain just the regression coefficients, their standard errors, and associated t statistics (thus omitting the additional information on the fit of the model provided by summary(jour_lm)), the function call coeftest(jour_lm) may be used. This has the advantage that other covariance matrix estimates—specified either as a function or directly as the fitted matrix—can be passed as an argument. For the journals data, this gives

R> coeftest(jour_lm, vcov = vcovHC)

t test of coefficients:

```
                Estimate Std. Error t value Pr(>|t|)
(Intercept)       4.7662     0.0555    85.8   <2e-16
log(citeprice)   -0.5331     0.0345   -15.5   <2e-16
```

which is almost identical to the standard summary() output. In the call above, the argument vcov was specified as a *function*; using coeftest(jour_lm, vcov = vcovHC(jour_lm)) (i.e., specifying vcov as the estimated covariance *matrix*) yields identical output.

To compare the different types of HC estimators described above for the journals data, we compute the corresponding standard errors for the fitted jour_lm model

R> t(sapply(c("const", "HC0", "HC1", "HC2", "HC3", "HC4"),
+   function(x) sqrt(diag(vcovHC(jour_lm, type = x)))))

```
      (Intercept) log(citeprice)
const     0.05591        0.03561
HC0       0.05495        0.03377
HC1       0.05526        0.03396
HC2       0.05525        0.03412
HC3       0.05555        0.03447
HC4       0.05536        0.03459
```

which shows that, for these data, all estimators lead to almost identical results. In fact, all standard errors are slightly smaller than those computed under the assumption of homoskedastic errors.

To illustrate that using robust covariance matrix estimates can indeed make a big difference, we reconsider the public schools data. We have already fitted a linear regression model explaining the per capita public school expenditures by per capita income and detected several influential observations (most notably Alaska). This has the effect that if a quadratic model is fitted to the data, it appears to provide a significant improvement over the linear model, although this spurious significance is in fact only caused by a single outlier, Alaska:

```
R> ps_lm <- lm(Expenditure ~ Income, data = ps)
R> ps_lm2 <- lm(Expenditure ~ Income + I(Income^2), data = ps)
R> anova(ps_lm, ps_lm2)

Analysis of Variance Table

Model 1: Expenditure ~ Income
Model 2: Expenditure ~ Income + I(Income^2)
  Res.Df    RSS Df Sum of Sq    F Pr(>F)
1     48 181015
2     47 150986  1     30030 9.35 0.0037
```

As illustrated by Cribari-Neto (2004), this can be remedied by using a robust covariance matrix estimate for inference. Note that in this context the pattern in the residuals is typically not referred to as heteroskedasticity but rather as having outliers or influential observations. However, the principles underlying the HC estimators also yield appropriate results in this situation.

To obtain the same type of test as in the `anova()` call above, we use `waldtest()`. This function provides a vcov argument, which again takes either a function or a matrix. In the latter case, the covariance matrix has to be computed for the more complex model to yield the correct result:

```
R> waldtest(ps_lm, ps_lm2, vcov = vcovHC(ps_lm2, type = "HC4"))

Wald test

Model 1: Expenditure ~ Income
Model 2: Expenditure ~ Income + I(Income^2)
  Res.Df Df    F Pr(>F)
1     48
2     47  1 0.08   0.77
```

This shows that the quadratic term is, in fact, not significant; an equivalent result can also be obtained via `coeftest(ps_lm2, vcov = vcovHC(ps_lm2, type = "HC4"))`.

HAC estimators

In time series regressions, if the error terms ε_i are correlated, Ω is not diagonal and can only be estimated directly upon introducing further assumptions on its structure. However, if the form of heteroskedasticity and autocorrelation is unknown, valid standard errors and tests may be obtained by employing empirical counterparts of $X^{\top}\Omega X$ instead. This is achieved by computing weighted sums of the empirical autocorrelations of $\hat{\varepsilon}_i x_i$.

Various estimators that differ with respect to the choice of the weights have been suggested. They have in common that weights are chosen according to some kernel function but differ in the choice of the kernel as well as the choice of the bandwidth used. The function vcovHAC() implements a general framework for this type of estimator; more details on its implementation can be found in Zeileis (2004). In addition to vcovHAC(), there are several convenience functions available in **sandwich** that make different strategies for choosing kernels and bandwidths easily accessible: the function NeweyWest() implements the estimator of Newey and West (1987, 1994) using a nonparametric bandwidth selection, kernHAC() provides the class of kernel HAC estimators of Andrews (1991) with parametric bandwidth selection as well as prewhitening, as suggested by Andrews and Monahan (1992), and weave() implements the class of weighted empirical adaptive variance estimators of Lumley and Heagerty (1999). In econometrics, the former two estimators are frequently used, and here we illustrate how they can be applied to the consumption function consump1.

We compute the standard errors under the assumption of spherical errors and compare them with the results of the quadratic spectral kernel and the Bartlett kernel HAC estimators, both using prewhitening:

```
R> rbind(SE = sqrt(diag(vcov(consump1))),
+    QS = sqrt(diag(kernHAC(consump1))),
+    NW = sqrt(diag(NeweyWest(consump1))))
```

```
   (Intercept)    dpi L(dpi)
SE       14.51 0.2063 0.2075
QS       94.11 0.3893 0.3669
NW      100.83 0.4230 0.3989
```

Both sets of robust standard errors are rather similar (except maybe for the intercept) and much larger than the uncorrected standard errors. As already illustrated above, these functions may again be passed to coeftest() or waldtest() (and other inference functions).

4.4 Resistant Regression

Leave-one-out diagnostics, as presented in Section 4.1, are a popular means for detecting unusual observations. However, such observations may "mask" each other, meaning that there may be several cases of the same type, and hence conventional diagnostics will not be able to detect such problems. With low-dimensional data such as PublicSchools, we can always resort to plotting, but the situation is much worse with high-dimensional data. A solution is to use robust regression techniques.

There are many procedures in statistics and econometrics labeled as "robust", among them the sandwich covariance estimators discussed in the preceding section. In this section, "robust" means "resistant" regression; that is, regression methods that can withstand alterations of a small percentage of the data set (in more practical terms, the estimates are unaffected by a certain percentage of outlying observations). The first regression estimators with this property, the least median of squares (LMS) and least trimmed squares (LTS) estimators defined by $\arg\min_\beta \text{med}_i |y_i - x_i^\top \beta|$ and $\arg\min_\beta \sum_{i=1}^q \hat{\varepsilon}_{i:n}^2(\beta)$, respectively (Rousseeuw 1984), were studied in the early 1980s. Here, $\hat{\varepsilon}_i^2 = |y_i - x_i^\top \beta|^2$ are the squared residuals and $i : n$ denotes that they are arranged in increasing order. It turned out that LTS is preferable, and we shall use this method below. For an applied account of robust regression techniques, we refer to Venables and Ripley (2002).

In econometrics, these methods appear to be not as widely used as they deserve to be. Zaman, Rousseeuw, and Orhan (2001) report that a search through ECONLIT "led to a remarkably low total of 14 papers" published in the economics literature that rely on robust regression techniques. This section is an attempt at improving the situation.

For illustration, we consider a growth regression. The determinants of economic growth have been a popular field of empirical research for the last 20 years. The starting point is the classical Solow model relating GDP growth to the accumulation of physical capital and to population growth. In an influential paper, Mankiw, Romer, and Weil (1992) consider an augmented Solow model that also includes a proxy for the accumulation of human capital. For a data set comprising 98 countries (available as GrowthDJ in **AER**), they conclude that human capital is an important determinant of economic growth. They also note that the OECD countries are not fitted well by their model. This led Nonneman and Vanhoudt (1996) to reexamine their findings for OECD countries (available as OECDGrowth in **AER**), suggesting, among other things, a further extension of the Solow model by incorporating a measure of accumulation of technological know-how.

Here, we consider the classical textbook Solow model for the Nonneman and Vanhoudt (1996) data with

$$\log(Y_t/Y_0) = \beta_1 + \beta_2 \log(Y_0) + \beta_3 \log(K_t) + \beta_4 \log(n_t + 0.05) + \varepsilon_t,$$

where Y_t and Y_0 are real GDP in periods t and 0, respectively, K_t is a measure of the accumulation of physical capital, and n_t is population growth (with 5% added in order to account for labor productivity growth). The OECDGrowth data provide these variables for all OECD countries with a population exceeding 1 million. Specifically, Y_t and Y_0 are taken as the real GDP (per person of working age; i.e., ages 15 to 65) in 1985 and 1960, respectively, both in 1985 international prices, K_t is the average of the annual ratios of real domestic investment to real GDP, and n_t is annual population growth, all for the period 1960–1985.

OLS estimation of this model yields

```
R> data("OECDGrowth")
R> solow_lm <- lm(log(gdp85/gdp60) ~ log(gdp60) +
+     log(invest) + log(popgrowth + .05), data = OECDGrowth)
R> summary(solow_lm)

Call:
lm(formula = log(gdp85/gdp60) ~ log(gdp60) + log(invest) +
   log(popgrowth + 0.05), data = OECDGrowth)

Residuals:
     Min       1Q   Median       3Q      Max
-0.18400 -0.03989 -0.00785  0.04506  0.31879

Coefficients:
                       Estimate Std. Error t value Pr(>|t|)
(Intercept)              2.9759     1.0216    2.91   0.0093
log(gdp60)              -0.3429     0.0565   -6.07 9.8e-06
log(invest)              0.6501     0.2020    3.22   0.0048
log(popgrowth + 0.05)   -0.5730     0.2904   -1.97   0.0640

Residual standard error: 0.133 on 18 degrees of freedom Multiple
R-squared: 0.746,              Adjusted R-squared: 0.704
F-statistic: 17.7 on 3 and 18 DF,  p-value: 1.34e-05
```

The fit is quite reasonable for a cross-section regression. The coefficients on gdp60 and invest are highly significant, and the coefficient on popgrowth is borderline at the 10% level. We shall return to this issue below.

With three regressors, the standard graphical displays are not as effective for the detection of unusual data. Zaman *et al.* (2001) recommend first running an LTS analysis flagging observations with unusually large residuals and running a standard OLS regression excluding the outlying observations thereafter. Since robust methods are not very common in applied econometrics, relying on OLS for the final estimates to be reported would seem to be a useful strategy for practical work. However, LTS may flag too many points as outlying. Recall from Section 4.1 that there are good and bad leverage points; only the bad ones should be excluded. Also, a large residual may correspond to an observation with small leverage, an observation with an unusual y_i that is not fitted well but at the same time does not disturb the analysis. The strategy is therefore to exclude observations that are bad leverage points, defined here as high-leverage points with large LTS residuals.

Least trimmed squares regression is provided in the function lqs() in the **MASS** package, the package accompanying Venables and Ripley (2002). Here, `lqs` stands for least quantile of squares because `lqs()` also provides LMS regression and generalizations thereof. However, the default is LTS, which is what we use here.

The code chunk

```
R> library("MASS")
R> solow_lts <- lqs(log(gdp85/gdp60) ~ log(gdp60) +
+    log(invest) + log(popgrowth + .05), data = OECDGrowth,
+    psamp = 13, nsamp = "exact")
```

sets `psamp = 13`, and thus we trim 9 of the 22 observations in `OECDGrowth`. By choosing `nsamp = "exact"`, the LTS estimates are computed by minimizing the sum of squares from all conceivable subsamples of size 13. This is only feasible for small samples such as the data under investigation; otherwise some other sampling technique should be used.[1]

`lqs()` provides two estimates of scale, the first defined via the fit criterion and the second based on the variance of those residuals whose absolute value is less than 2.5 times the initial estimate. Following Zaman *et al.* (2001), we use the second estimate of scale and define a large residual as a scaled residual exceeding 2.5 in absolute values. Thus, the observations corresponding to small residuals can be extracted by

```
R> smallresid <- which(
+    abs(residuals(solow_lts)/solow_lts$scale[2]) <= 2.5)
```

We still need a method for detecting the high-leverage points. For consistency, this method should be a robust method. The robustness literature provides several robust covariance estimators that can be used to determine these points, among them the minimum-volume ellipsoid (MVE) and the minimum-covariance determinant (MCD) methods. Both are implemented in the function `cov.rob()` from the **MASS** package, with MVE being the default.

Below, we extract the model matrix, estimate its covariance matrix by MVE, and subsequently compute the leverage (utilizing the `mahalanobis()` function), storing the observations that are not high-leverage points.

```
R> X <- model.matrix(solow_lm)[,-1]
R> Xcv <- cov.rob(X, nsamp = "exact")
R> nohighlev <- which(
+    sqrt(mahalanobis(X, Xcv$center, Xcv$cov)) <= 2.5)
```

The "good observations" are defined as those having at least one of the desired properties, small residual or low leverage. They are determined by concatenating the vectors `smallresid` and `nohighlev` and removing duplicates using `unique()`:

```
R> goodobs <- unique(c(smallresid, nohighlev))
```

---

[1] Our results slightly improve on Zaman *et al.* (2001) because they do not seem to have used an exhaustive search for determining their robust estimates.

Thus, the "bad observations" are

```
R> rownames(OECDGrowth)[-goodobs]
```

```
[1] "Canada"    "USA"       "Turkey"    "Australia"
```

Running OLS excluding the bad leverage points now yields

```
R> solow_rob <- update(solow_lm, subset = goodobs)
R> summary(solow_rob)

Call:
lm(formula = log(gdp85/gdp60) ~ log(gdp60) + log(invest) +
    log(popgrowth + 0.05), data = OECDGrowth,
    subset = goodobs)

Residuals:
    Min      1Q  Median      3Q     Max
-0.1545 -0.0555 -0.0065  0.0316  0.2677

Coefficients:
                       Estimate Std. Error t value Pr(>|t|)
(Intercept)              3.7764     1.2816    2.95   0.0106
log(gdp60)              -0.4507     0.0569   -7.93  1.5e-06
log(invest)              0.7033     0.1906    3.69   0.0024
log(popgrowth + 0.05)   -0.6504     0.4190   -1.55   0.1429

Residual standard error: 0.107 on 14 degrees of freedom
Multiple R-squared: 0.853,          Adjusted R-squared: 0.822
F-statistic: 27.1 on 3 and 14 DF,  p-value: 4.3e-06
```

Note that the results are somewhat different from the original OLS results for the full data set. Specifically, population growth does not seem to belong in this model. Of course, this does not mean that population growth plays no role in connection with economic growth but just that this variable is not needed conditional on the inclusion of the remaining ones and, more importantly, for this subset of countries. With a larger set of countries, population growth is quite likely to play its role. The OECD countries are fairly homogeneous with respect to that variable, and some countries with substantial population growth have been excluded in the robust fit. Hence, the result should not come as a surprise.

Augmented or extended versions of the Solow model that include further regressors such as human capital (log(school)) and technological know-how (log(randd)) are explored in an exercise.

4.5 Quantile Regression

Least-squares regression can be viewed as a method for modeling the conditional mean of a response. Sometimes other characteristics of the conditional distribution are more interesting, for example the median or more generally the quantiles. Thanks to the efforts of Roger Koenker and his co-authors, quantile regression has recently been gaining ground as an alternative to OLS in many econometric applications; see Koenker and Hallock (2001) for a brief introduction and Koenker (2005) for a comprehensive treatment.

The (linear) quantile regression model is given by the conditional quantile functions (indexed by the quantile τ)

$$Q_y(\tau|x) = x_i^\top \beta;$$

i.e., $Q_y(\tau|x)$ denotes the τ-quantile of y conditional on x. Estimates are obtained by minimizing $\sum_i \varrho_\tau(y_i - x_i^\top \beta)$ with respect to β, where, for $\tau \in (0,1)$, ϱ_τ denotes the piecewise linear function $\varrho_\tau(u) = u\{\tau - I(u < 0)\}$, I being the indicator function. This is a linear programming problem.

A fitting function, $\mathtt{rq()}$, for "regression quantiles", has long been available in the package **quantreg** (Koenker 2008). For a brief illustration, we return to the Bierens and Ginther (2001) data used in Chapter 3 and consider quantile versions of a Mincer-type wage equation, namely

$$Q_{\log(\mathtt{wage})}(\tau|x) = \beta_1 + \beta_2\,\mathtt{experience} + \beta_3\,\mathtt{experience}^2 + \beta_4\,\mathtt{education}$$

The function $\mathtt{rq()}$ defaults to $\tau = 0.5$; i.e., median or LAD (for "least absolute deviations") regression. Hence a median version of the wage equation is fitted via

```
R> library("quantreg")
R> data("CPS1988")
R> cps_f <- log(wage) ~ experience + I(experience^2) + education
R> cps_lad <- rq(cps_f, data = CPS1988)
R> summary(cps_lad)

Call: rq(formula = cps_f, data = CPS1988)


tau: [1] 0.5


Coefficients:
                   Value     Std. Error t value    Pr(>|t|)
(Intercept)        4.24088   0.02190    193.67801  0.00000
experience         0.07744   0.00115    67.50040   0.00000
I(experience^2)   -0.00130   0.00003   -49.97890   0.00000
education          0.09429   0.00140    67.57170   0.00000
```

This may be compared with the OLS results given in the preceding chapter.

Quantile regression is particularly useful when modeling several quantiles simultaneously. In order to illustrate some basic functions from **quantreg**, we consider the first and third quartiles (i.e., $\tau = 0.25$ and $\tau = 0.75$). Since rq() takes vectors of quantiles, fitting these two models is as easy as

```
R> cps_rq <- rq(cps_f, tau = c(0.25, 0.75), data = CPS1988)
R> summary(cps_rq)

Call: rq(formula = cps_f, tau = c(0.25, 0.75), data = CPS1988)

tau: [1] 0.25

Coefficients:
                 Value    Std. Error t value    Pr(>|t|)
(Intercept)       3.78227  0.02866   131.95187   0.00000
experience        0.09156  0.00152    60.26473   0.00000
I(experience^2)  -0.00164  0.00004   -45.39064   0.00000
education         0.09321  0.00185    50.32519   0.00000

Call: rq(formula = cps_f, tau = c(0.25, 0.75), data = CPS1988)

tau: [1] 0.75

Coefficients:
                 Value    Std. Error t value    Pr(>|t|)
(Intercept)       4.66005  0.02023   230.39729   0.00000
experience        0.06377  0.00097    65.41363   0.00000
I(experience^2)  -0.00099  0.00002   -44.15591   0.00000
education         0.09434  0.00134    70.65853   0.00000
```

A natural question is whether the regression lines or surfaces are parallel; i.e., whether the effects of the regressors are uniform across quantiles. There exists an anova() method for exploring this question. It requires separate fits for each quantile and can be used in two forms: for an overall test of equality of the entire sets of coefficients, we use

```
R> cps_rq25 <- rq(cps_f, tau = 0.25, data = CPS1988)
R> cps_rq75 <- rq(cps_f, tau = 0.75, data = CPS1988)
R> anova(cps_rq25, cps_rq75)

Quantile Regression Analysis of Variance Table

Model: log(wage) ~ experience + I(experience^2) + education
Joint Test of Equality of Slopes: tau in {  0.25 0.75  }

     Df Resid Df F value Pr(>F)
1     3    56307     115 <2e-16
```

while

```
R> anova(cps_rq25, cps_rq75, joint = FALSE)
```

Quantile Regression Analysis of Variance Table

Model: log(wage) ~ experience + I(experience^2) + education
Tests of Equality of Distinct Slopes: tau in { 0.25 0.75 }

	Df	Resid Df	F value	Pr(>F)
experience	1	56309	339.41	<2e-16
I(experience^2)	1	56309	329.74	<2e-16
education	1	56309	0.35	0.55

provides coefficient-wise comparisons. We see that effects are not uniform across quantiles in this example, with differences being associated with the regressor experience.

It is illuminating to visualize the results from quantile regression fits. One possibility is to plot, for each regressor, the estimate as a function of the quantile. This is achieved using plot() on the summary() of the quantile regression object. In order to obtain a more meaningful plot, we now use a larger set of τs, specifically $\tau \in \{0.05, 0.1, \ldots, 0.95\}$:

```
R> cps_rqbig <- rq(cps_f, tau = seq(0.05, 0.95, by = 0.05),
+     data = CPS1988)
R> cps_rqbigs <- summary(cps_rqbig)
```

Figure 4.5, obtained via

```
R> plot(cps_rqbigs)
```

visualizes the variation of the coefficients as a function of τ, and it is clear that the influence of the covariates is far from uniform. The shaded areas represent pointwise 90% (by default) confidence intervals for the quantile regression estimates. For comparison, the horizontal solid and dashed lines shown in each plot signify the OLS estimate and an associated 90% confidence interval.

It should be noted that **quantreg** contains a number of further functions for quantile modeling, including nonlinear and nonparametric versions. There also exist several algorithms for fitting these models (specifically, both exterior and interior point methods) as well as several choices of methods for computing confidence intervals and related test statistics.

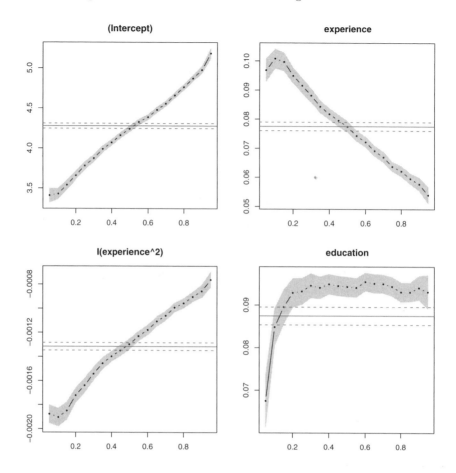

Fig. 4.5. Visualization of a quantile regression fit.

4.6 Exercises

1. Consider the `CigarettesB` data taken from Baltagi (2002). Run a regression of real per capita consumption on real price and real per capita income (all variables in logarithms). Obtain the usual diagnostic statistics using `influence.measures()`. Which observations are influential? To which states do they correspond? Are the results intuitive?
2. Reanalyze the `PublicSchools` data using robust methods:
    (a) Run a regression of `Expenditure` on `Income` using least trimmed squares (LTS). Which observations correspond to large LTS residuals?
    (b) Which observations are high-leverage points?

(c) Run OLS on the data, excluding all observations that have large LTS residuals and are high-leverage points. Compare your result with the analysis provided in Section 4.1.

3. Explore further growth regressions for the OECDGrowth data using the augmented and extended Solow models of Nonneman and Vanhoudt (1996), which consider the additional regressors log(school) (human capital) and log(randd) (technological know-how), respectively. First, replicate the OLS results from Nonneman and Vanhoudt (1996, Table IV), and subsequently compare them with the resistant LTS results by adopting the strategy of Zaman et al. (2001).

4. When discussing quantile regression, we confined ourselves to the standard Mincer equation. However, the CPS1988 data contain further explanatory variables, namely the factors ethnicity, smsa, region, and parttime. Replicate the LAD regression (i.e., the quantile regression with $\tau = 0.5$) results from Bierens and Ginther (2001) using these covariates.

5

Models of Microeconometrics

Many econometrics packages can perform the analyses discussed in the following sections. Often, however, they do so with a different program or procedure for each type of analysis—for example, probit regression and Poisson regression—so that the unifying structure of these methods is not apparent.

R does not come with different programs for probit and Poisson regressions. Instead, it follows mainstream statistics in providing the unifying framework of generalized linear models (GLMs) and a single fitting function, `glm()`. Furthermore, models extending GLMs are provided by R functions that analogously extend the basic `glm()` function (i.e., have similar interfaces, return values, and associated methods).

This chapter begins with a brief introduction to GLMs, followed by sections on regression for binary dependent variables, counts, and censored dependent variables. The final section points to methods for multinomial and ordinal responses and semiparametric extensions.

5.1 Generalized Linear Models

Chapter 3 was devoted to the linear regression model, for which inference is exact and the OLS estimator coincides with the maximum likelihood estimator (MLE) when the disturbances are, conditional on the regressors, i.i.d. $\mathcal{N}(0, \sigma^2)$. Here, we briefly describe how the salient features of this model can be extended to situations where the dependent variable y comes from a wider class of distributions.

Three aspects of the linear regression model for a conditionally normally distributed response y are:

1. The linear predictor $\eta_i = x_i^\top \beta$ through which $\mu_i = \mathsf{E}(y_i|x_i)$ depends on the $k \times 1$ vectors x_i of observations and β of parameters.
2. The distribution of the dependent variable $y_i|x_i$ is $\mathcal{N}(\mu_i, \sigma^2)$.
3. The expected response is equal to the linear predictor, $\mu_i = \eta_i$.

C. Kleiber, A. Zeileis, *Applied Econometrics with R*,
DOI: 10.1007/978-0-387-77318-6_5, © Springer Science+Business Media, LLC 2008

The class of generalized linear models (GLMs) extends 2. and 3. to more general families of distributions for y and to more general relations between $\mathsf{E}(y_i|x_i)$ and the linear predictor than the identity. Specifically, $y_i|x_i$ may now follow a density or probability mass function of the type

$$f(y;\theta,\phi) = \exp\left\{\frac{y\theta - b(\theta)}{\phi} + c(y;\phi)\right\}, \tag{5.1}$$

where θ, called the canonical parameter, depends on the linear predictor, and the additional parameter ϕ, called the dispersion parameter, is often known. Also, the linear predictor and the expectation of y are now related by a monotonic transformation,

$$g(\mu_i) = \eta_i.$$

For fixed ϕ, (5.1) describes a linear exponential family, a class of distributions that includes a number of well-known distributions such as the normal, Poisson, and binomial.

The class of generalized linear models is thus defined by the following elements:

1. The linear predictor $\eta_i = x_i^\top \beta$ through which $\mu_i = \mathsf{E}(y_i|x_i)$ depends on the $k \times 1$ vectors x_i of observations and β of parameters.
2. The distribution of the dependent variable $y_i|x_i$ is a linear exponential family.
3. The expected response and the linear predictor are related by a monotonic transformation, $g(\mu_i) = \eta_i$, called the link function of the GLM.

Thus, the family of GLMs extends the applicability of linear-model ideas to data where responses are binary or counts, among further possibilities. The unifying framework of GLMs emerged in the statistical literature in the early 1970s (Nelder and Wedderburn 1972).

The Poisson distribution with parameter μ and probability mass function

$$f(y;\mu) = \frac{e^{-\mu}\mu^y}{y!}, \qquad y = 0, 1, 2, \ldots,$$

perhaps provides the simplest example leading to a nonnormal GLM. Writing

$$f(y;\mu) = \exp(y\log\mu - \mu - \log y!),$$

it follows that the Poisson density has the form (5.1) with $\theta = \log\mu$, $b(\theta) = e^\theta$, $\phi = 1$, and $c(y;\phi) = -\log y!$. Furthermore, in view of $\mathsf{E}(y) = \mu > 0$, it is natural to employ $\log\mu = \eta$; i.e., to use a logarithmic link. The transformation g relating the original parameter, here μ, and the canonical parameter θ from the exponential family representation is called the canonical link in the GLM literature. Hence the logarithmic link is in fact the canonical link for the Poisson family.

Table 5.1. Selected GLM families and their canonical (default) links.

Family	Canonical link	Name
binomial	$\log\{\mu/(1-\mu)\}$	logit
gaussian	μ	identity
poisson	$\log\mu$	log

Similarly, for binary data, the Bernoulli distribution (a special case of the binomial distribution) is suitable, and here we have

$$f(y;p) = \left\{ y \log \left(\frac{p}{1-p} \right) + \log(1-p) \right\}, \quad y \in \{0,1\},$$

which shows that the logit transform $\log\{p/(1-p)\}$, the quantile function of the logistic distribution, corresponds to the canonical link for the binary regression model. A widely used non-canonical link for this model is the quantile function of the standard normal distribution, yielding the probit link.

Table 5.1 provides a summary of the most important examples of generalized linear models. A more complete list may be found in standard references on GLMs such as McCullagh and Nelder (1989).

In view of the built-in distributional assumption, it is natural to estimate a GLM by the method of maximum likelihood. For the GLM with a normally distributed response and the identity link, the MLE reduces to the least-squares estimator and is therefore available in closed form. In general, there is no closed-form expression for the MLE; instead it must be determined using numerical methods. For GLMs, the standard algorithm is called iterative weighted least squares (IWLS). It is an implementation of the familiar Fisher scoring algorithm adapted for GLMs.

These analogies with the classical linear model suggest that a fitting function for GLMs could look almost like the fitting function for the linear model, lm(). In R, this is indeed the case. The fitting function for GLMs is called glm(), and its syntax closely resembles the syntax of lm(). Of course, there are extra arguments for selecting the response distribution and the link function, but apart from this, the familiar arguments such as formula, data, weights, and subset are all available. In addition, all the extractor functions from Table 3.1 have methods for objects of class "glm", the class of objects returned by glm().

5.2 Binary Dependent Variables

Regression problems with binary dependent variables are quite common in microeconometrics, usually under the names of logit and probit regressions. The model is

$$E(y_i|x_i) = p_i = F(x_i^\top \beta), \quad i = 1, \ldots, n,$$

where F equals the standard normal CDF in the probit case and the logistic CDF in the logit case. As noted above, fitting logit or probit models proceeds using the function glm() with the appropriate family argument (including a specification of the link function). For binary responses (i.e., Bernoulli outcomes), the family is binomial, and the link is specified either as link = "logit" or link = "probit", the former being the default. (A look at ?glm reveals that there are further link functions available, but these are not commonly used in econometrics.)

To provide a typical example, we again turn to labor economics, considering female labor force participation for a sample of 872 women from Switzerland. The data were originally analyzed by Gerfin (1996) and are also used by Davidson and MacKinnon (2004). The dependent variable is participation, which we regress on all further variables plus age squared; i.e., on income, education, age, age^2, numbers of younger and older children (youngkids and oldkids), and on the factor foreign, which indicates citizenship. Following Gerfin (1996), we begin with a probit regression:

```
R> data("SwissLabor")
R> swiss_probit <- glm(participation ~ . + I(age^2),
+     data = SwissLabor, family = binomial(link = "probit"))
R> summary(swiss_probit)

Call:
glm(formula = participation ~ . + I(age^2),
    family = binomial(link = "probit"), data = SwissLabor)

Deviance Residuals:
   Min      1Q  Median      3Q     Max
-1.919  -0.969  -0.479   1.021   2.480

Coefficients:
              Estimate Std. Error z value Pr(>|z|)
(Intercept)     3.7491     1.4069    2.66   0.0077
income         -0.6669     0.1320   -5.05  4.3e-07
age             2.0753     0.4054    5.12  3.1e-07
education       0.0192     0.0179    1.07   0.2843
youngkids      -0.7145     0.1004   -7.12  1.1e-12
oldkids        -0.1470     0.0509   -2.89   0.0039
foreignyes      0.7144     0.1213    5.89  3.9e-09
I(age^2)       -0.2943     0.0499   -5.89  3.8e-09

(Dispersion parameter for binomial family taken to be 1)

    Null deviance: 1203.2  on 871  degrees of freedom
```

Residual deviance: 1017.2 on 864 degrees of freedom
AIC: 1033

Number of Fisher Scoring iterations: 4

This shows that the summary of a "glm" object closely resembles the summary of an "lm" object: there are a brief summary of the residuals and a table providing coefficients, standard errors, etc., and this is followed by some further summary measures. Among the differences is the column labeled z value. It provides the familiar t statistic (the estimate divided by its standard error), but since it does not follow an exact t distribution here, even under ideal conditions, it is commonly referred to its asymptotic approximation, the normal distribution. Hence, the p value comes from the standard normal distribution here. For the current model, all variables except education are highly significant. The dispersion parameter (denoted ϕ in (5.1)) is taken to be 1 because the binomial distribution is a one-parameter exponential family. The deviance resembles the familiar residual sum of squares. Finally, the number of Fisher scoring iterations is provided, which indicates how quickly the IWLS algorithm terminates.

Visualization

Traditional scatterplots are of limited use in connection with binary dependent variables. When plotting participation versus a continuous regressor such as age using

```R
R> plot(participation ~ age, data = SwissLabor, ylevels = 2:1)
```

R by default provides a so-called spinogram, which is similar to the spine plot introduced in Section 2.8. It first groups the regressor age into intervals, just as in a histogram, and then produces a spine plot for the resulting proportions of participation within the age groups. Note that the horizontal axis is distorted because the width of each age interval is proportional to the corresponding number of observations. By setting ylevels = 2:1, the order of participation levels is reversed, highlighting participation (rather than non-participation). Figure 5.1 shows the resulting plot for the regressors education and age, indicating an approximately quadratic relationship between participation and age and slight nonlinearities between participation and education.

Effects

In view of

$$\frac{\partial \mathsf{E}(y_i|x_i)}{\partial x_{ij}} = \frac{\partial \Phi(x_i^\top \beta)}{\partial x_{ij}} = \phi(x_i^\top \beta) \cdot \beta_j, \tag{5.2}$$

effects in a probit regression model vary with the regressors. Researchers often state average marginal effects when reporting results from a binary

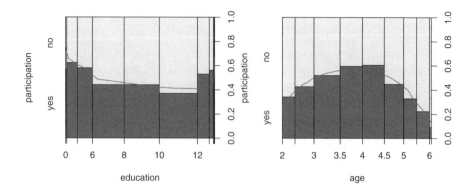

Fig. 5.1. Spinograms for binary dependent variables.

GLM. There are several versions of such averages. It is perhaps best to use $n^{-1} \sum_{i=1}^{n} \phi(x_i^\top \hat{\beta}) \cdot \hat{\beta}_j$, the average of the sample marginal effects. In R, this is simply

```
R> fav <- mean(dnorm(predict(swiss_probit, type = "link")))
R> fav * coef(swiss_probit)
```

```
(Intercept)      income         age   education    youngkids
   1.241930   -0.220932    0.687466    0.006359    -0.236682
    oldkids  foreignyes    I(age^2)
  -0.048690    0.236644   -0.097505
```

Another version evaluates (5.2) at the average regressor, $\bar{x}$. This is straightforward as long as all regressors are continuous; however, more often than not, the model includes factors. In this case, it is preferable to report average effects for all levels of the factors, averaging only over continuous regressors. In the case of the SwissLabor data, there is only a single factor, foreign, indicating whether the individual is Swiss or not. Thus, average marginal effects for these groups are available via

```
R> av <- colMeans(SwissLabor[, -c(1, 7)])
R> av <- data.frame(rbind(swiss = av, foreign = av),
+    foreign = factor(c("no", "yes")))
R> av <- predict(swiss_probit, newdata = av, type = "link")
R> av <- dnorm(av)
R> av["swiss"] * coef(swiss_probit)[-7]
```

```
(Intercept)      income         age   education    youngkids
   1.495137   -0.265976    0.827628    0.007655    -0.284938
```

```
    oldkids      I(age^2)
   -0.058617    -0.117384
```

and

```
R> av["foreign"] * coef(swiss_probit)[-7]
```

```
(Intercept)       income        age    education    youngkids
  1.136517     -0.202180   0.629115     0.005819    -0.216593
    oldkids      I(age^2)
  -0.044557    -0.089229
```

This indicates that all effects are smaller in absolute size for the foreign women.

Furthermore, methods for visualizing effects in linear and generalized linear models are available in the package **effects** (Fox 2003).

Goodness of fit and prediction

In contrast to the linear regression model, there is no commonly accepted version of R^2 for generalized linear models, not even for the special case of binary dependent variables.

R^2 measures for nonlinear models are often called pseudo-R^2s. If we define $\ell(\hat{\beta})$ as the log-likelihood for the fitted model and $\ell(\bar{y})$ as the log-likelihood for the model containing only a constant term, we can use a ratio of log-likelihoods,

$$R^2 = 1 - \frac{\ell(\hat{\beta})}{\ell(\bar{y})},$$

often called McFadden's pseudo-R^2. This is easily made available in R by computing the null model, and then extracting the `logLik()` values for the two models

```
R> swiss_probit0 <- update(swiss_probit, formula = . ~ 1)
R> 1 - as.vector(logLik(swiss_probit)/logLik(swiss_probit0))
```

```
[1] 0.1546
```

yields a rather modest pseudo-R^2 for the model fitted to the `SwissLabor` data. In the first line of the code chunk above, the `update()` call reevaluates the original call using the formula corresponding to the null model. In the second line, `as.vector()` strips off all additional attributes returned by `logLik()` (namely the `"class"` and `"df"` attributes).

A further way of assessing the fit of a binary regression model is to compare the categories of the observed responses with their fitted values. This requires prediction for GLMs. The generic function `predict()` also has a method for objects of class "glm". However, for GLMs, things are slightly more involved than for linear models in that there are several types of predictions. Specifically, the `predict()` method for "glm" objects has a **type** argument allowing types "link", "response", and "terms". The default "link" is on the scale of the linear predictors, and "response" is on the scale of the mean of the response variable. For a binary regression model, the default predictions are thus of probabilities on the logit or probit scales, while type = "response" yields the predicted probabilities themselves. (In addition, the "terms" option returns a matrix giving the fitted values of each term in the model formula on the linear-predictor scale.)

In order to obtain the predicted class, we round the predicted probabilities and tabulate the result against the actual values of participation in a confusion matrix,

```
R> table(true = SwissLabor$participation,
+     pred = round(fitted(swiss_probit)))

      pred
true    0    1
   no  337  134
   yes 146  255
```

corresponding to 67.89% correctly classified and 32.11% misclassified observations in the observed sample.

However, this evaluation of the model uses the arbitrarily chosen cutoff 0.5 for the predicted probabilities. To avoid choosing a particular cutoff, the performance can be evaluated for every conceivable cutoff; e.g., using (as above) the "accuracy" of the model, the proportion of correctly classified observations, as the performance measure. The left panel of Figure 5.2 indicates that the best accuracy is achieved for a cutoff slightly below 0.5.

Alternatively, the receiver operating characteristic (ROC) curve can be used: for every cutoff $c \in [0, 1]$, the associated true positive rate (TPR(c), in our case the number of women participating in the labor force that are also classified as participating compared with the total number of women participating) is plotted against the false positive rate (FPR(c), in our case the number of women not participating in the labor force that are classified as participating compared with the total number of women not participating). Thus, ROC $= \{(\text{FPR}(c), \text{TPR}(c)) \mid c \in [0, 1]\}$, and this curve is displayed in the right panel of Figure 5.2. For a sensible predictive model, the ROC curve should be at least above the diagonal (which corresponds to random guessing). The closer the curve is to the upper left corner (FPR $= 0$, TPR $= 1$), the better the model performs.

In R, visualizations of these (and many other performance measures) can be created using the **ROCR** package (Sing, Sander, Beerenwinkel, and Lengauer 2005). In the first step, the observations and predictions are captured in an object created by prediction(). Subsequently, various performances can be computed and plotted:

```
R> library("ROCR")
R> pred <- prediction(fitted(swiss_probit),
```

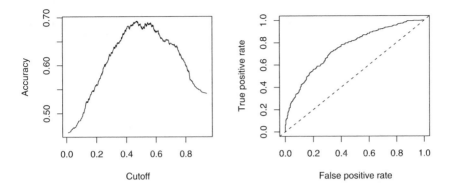

Fig. 5.2. Accuracy and ROC curve for labor force probit regression.

```
+    SwissLabor$participation)
R> plot(performance(pred, "acc"))
R> plot(performance(pred, "tpr", "fpr"))
R> abline(0, 1, lty = 2)
```

Figure 5.2 indicates that the fit is reasonable but also that there is room for improvement. We will reconsider this problem at the end of this chapter using semiparametric techniques.

Residuals and diagnostics

For residual-based diagnostics, a `residuals()` method for "glm" objects is available. It provides various types of residuals, the most prominent of which are deviance and Pearson residuals. The former are defined as (signed) contributions to the overall deviance of the model and are computed by default in R. The latter are the raw residuals $y_i - \hat{\mu}_i$ scaled by the standard error (often called standardized residuals in econometrics) and are available by setting `type = "pearson"`. Other types of residuals useful in certain situations and readily available via the `residuals()` method are working, raw (or response), and partial residuals. The associated sums of squares can be inspected using

```
R> deviance(swiss_probit)

[1] 1017

R> sum(residuals(swiss_probit, type = "deviance")^2)

[1] 1017

R> sum(residuals(swiss_probit, type = "pearson")^2)
```

```
[1] 866.5
```

and an analysis of deviance is performed by the `anova()` method for "glm" objects. Other standard tests for nested model comparisons, such as `waldtest()`, `linear.hypothesis()`, and `coeftest()`, are available as well.

We also note that sandwich estimates of the covariance matrix are available in the usual manner, and thus

```
R> coeftest(swiss_probit, vcov = sandwich)
```

would give the usual regression output with robustified standard errors and t statistics. However, the binary case differs from the linear regression model in that it is not possible to misspecify the variance while correctly specifying the regression equation. Instead, both are either correctly specified or not. Thus, users must be aware that, in the case where conventional and sandwich standard errors are very different, there are likely to be problems with the regression itself and not just with the variances of the estimates. Therefore, we do not recommend general use of sandwich estimates in the binary regression case (see also Freedman 2006, for further discussion). Sandwich estimates are much more useful and less controversial in Poisson regressions, as discussed in the following section.

(Quasi-)complete separation

To conclude this section, we briefly discuss an issue that occasionally arises with probit or logit regressions. For illustration, we consider a textbook example inspired by Stokes (2004). The goal is to study the deterrent effect of capital punishment in the United States of America in 1950 utilizing the `MurderRates` data taken from Maddala (2001). Maddala uses these data to illustrate probit and logit models, apparently without noticing that there is a problem.

Running a logit regression of an indicator of the incidence of executions (`executions`) during 1946–1950 on the median time served of convicted murderers released in 1951 (`time` in months), the median family income in 1949 (`income`), the labor force participation rate (in percent) in 1950 (`lfp`), the proportion of the population that was non-Caucasian in 1950 (`noncauc`), and a factor indicating region (`southern`) yields, using the defaults of the estimation process,

```
R> data("MurderRates")
R> murder_logit <- glm(I(executions > 0) ~ time + income +
+     noncauc + lfp + southern, data = MurderRates,
+     family = binomial)
```

```
Warning message:
fitted probabilities numerically 0 or 1 occurred in:
glm.fit(x = X, y = Y, weights = weights, start = start,
```

Thus, calling `glm()` results in a warning message according to which some fitted probabilities are numerically identical to zero or one. Also, the standard error of **southern** is suspiciously large:

```
R> coeftest(murder_logit)
```

z test of coefficients:

	Estimate	Std. Error	z value	Pr(>|z|)
(Intercept)	10.9933	20.7734	0.53	0.597
time	0.0194	0.0104	1.87	0.062
income	10.6101	5.6541	1.88	0.061
noncauc	70.9879	36.4118	1.95	0.051
lfp	-0.6676	0.4767	-1.40	0.161
southernyes	17.3313	2872.1707	0.01	0.995

Clearly, this model deserves a closer look. The warning suggests that numerical problems were encountered, so it is advisable to modify the default settings of the IWLS algorithm in order to determine the source of the phenomenon. The relevant argument to `glm()` is `control`, which takes a list consisting of the entries `epsilon`, the convergence tolerance epsilon, `maxit`, the maximum number of IWLS iterations, and `trace`, the latter indicating if intermediate output is required for each iteration. Simultaneously decreasing the epsilon and increasing the maximum number of iterations yields

```
R> murder_logit2 <- glm(I(executions > 0) ~ time + income +
+    noncauc + lfp + southern, data = MurderRates,
+    family = binomial, control = list(epsilon = 1e-15,
+    maxit = 50, trace = FALSE))
```

```
Warning message:
fitted probabilities numerically 0 or 1 occurred in:
glm.fit(x = X, y = Y, weights = weights, start = start,
```

Interestingly, the warning does not go away and the coefficient on **southern** has doubled, accompanied by a 6,000-fold increase of the corresponding standard error:

```
R> coeftest(murder_logit2)
```

z test of coefficients:

	Estimate	Std. Error	z value	Pr(>|z|)
(Intercept)	1.10e+01	2.08e+01	0.53	0.597
time	1.94e-02	1.04e-02	1.87	0.062
income	1.06e+01	5.65e+00	1.88	0.061
noncauc	7.10e+01	3.64e+01	1.95	0.051
lfp	-6.68e-01	4.77e-01	-1.40	0.161
southernyes	3.13e+01	1.73e+07	1.8e-06	1.000

The explanation of this phenomenon is somewhat technical: although the likelihood of the logit model is known to be globally concave and bounded from above, this does not imply that an interior maximum exists. This is precisely the problem we encountered here, and it depends on the settings of the function call when and where the algorithm terminates. Termination does not mean that a maximum was found, just that it was not possible to increase the objective function beyond the tolerance epsilon supplied. Specifically, we have here a situation where the maximum likelihood estimator does not exist. Instead, there exists a β^0 such that

$$y_i = 0 \quad \text{whenever} \quad x_i^\top \beta^0 \leq 0,$$
$$y_i = 1 \quad \text{whenever} \quad x_i^\top \beta^0 \geq 0.$$

If this is the case, the data are said to exhibit quasi-complete separation (the case of strict inequalities being called complete separation). Although the effect is not uncommon with small data sets, it is rarely discussed in textbooks; Davidson and MacKinnon (2004) is an exception.

For the problem at hand, the change in the coefficient on southern already indicates that this variable alone is responsible for the effect. A tabulation reveals

```
R> table(I(MurderRates$executions > 0), MurderRates$southern)
```

```
        no yes
FALSE   9   0
TRUE   20  15
```

In short, all of the 15 southern states, plus 20 of the remaining ones, executed convicted murderers during the period in question; thus the variable southern alone contains a lot of information on the dependent variable.

In practical terms, complete or quasi-complete separation is not necessarily a nuisance. After all, we are able to perfectly distinguish the zeros from the ones. However, many practitioners find it counterintuitive, if not disturbing, that the nonexistence of the MLE might be beneficial in some situations. Note also that inspection of the individual t statistics in coeftest(murder_logit) suggests excluding southern. As a result, the warning would have gone away, but the predictions would have been worse. The message is to carefully study such warnings in regressions with a binary dependent variable, with huge standard errors often pointing to the source of the problem.

5.3 Regression Models for Count Data

In this section, we consider a number of regression models for count data. A convenient reference on the methodological background is Cameron and Trivedi (1998). For illustration, we use the RecreationDemand data previously analyzed by Ozuna and Gomez (1995) and Gurmu and Trivedi (1996), among others. The data are cross-section data on the number of recreational boating trips to Lake Somerville, Texas, in 1980, based on a survey administered to 2,000 registered leisure boat owners in 23 counties in eastern Texas. The dependent variable is trips, and we want to regress it on all further variables: a (subjective) quality ranking of the facility (quality), a factor indicating whether the individual engaged in water-skiing at the lake (ski), household income (income), a factor indicating whether the individual paid a user's fee at the lake (userfee), and three cost variables (costC, costS, costH) representing opportunity costs.

We begin with the standard model for count data, a Poisson regression. As noted above, this is a generalized linear model. Using the canonical link for the Poisson family (the log link), the model is

$$E(y_i|x_i) = \mu_i = \exp(x_i^\top \beta).$$

Fitting is as simple as

```
R> data("RecreationDemand")
R> rd_pois <- glm(trips ~ ., data = RecreationDemand,
+    family = poisson)
```

To save space, we only present the partial coefficient tests and not the full summary() output:

```
R> coeftest(rd_pois)
```

```
z test of coefficients:
```

	Estimate	Std. Error	z value	Pr(>|z|)
(Intercept)	0.26499	0.09372	2.83	0.0047
quality	0.47173	0.01709	27.60	< 2e-16
skiyes	0.41821	0.05719	7.31	2.6e-13
income	-0.11132	0.01959	-5.68	1.3e-08
userfeeyes	0.89817	0.07899	11.37	< 2e-16
costC	-0.00343	0.00312	-1.10	0.2713
costS	-0.04254	0.00167	-25.47	< 2e-16
costH	0.03613	0.00271	13.34	< 2e-16

This would seem to indicate that almost all regressors are highly significant. For later reference, we note the log-likelihood of the fitted model:

```
R> logLik(rd_pois)
```

```
'log Lik.' -1529 (df=8)
```

Dealing with overdispersion

Recall that the Poisson distribution has the property that the variance equals the mean ("equidispersion"). This built-in feature needs to be checked in any empirical application. In econometrics, Poisson regressions are often plagued by overdispersion, meaning that the variance is larger than the linear predictor permits.

One way of testing for overdispersion is to consider the alternative hypothesis (Cameron and Trivedi 1990)

$$\mathsf{Var}(y_i|x_i) = \mu_i + \alpha \cdot h(\mu_i), \tag{5.3}$$

where h is a positive function of μ_i. Overdispersion corresponds to $\alpha > 0$ and underdispersion to $\alpha < 0$. The coefficient α can be estimated by an auxiliary OLS regression and tested with the corresponding t statistic, which is asymptotically standard normal under the null hypothesis of equidispersion.

Common specifications of the transformation function h are $h(\mu) = \mu^2$ or $h(\mu) = \mu$. The former corresponds to a negative binomial (NB) model (see below) with quadratic variance function (called NB2 by Cameron and Trivedi 1998), the latter to an NB model with linear variance function (called NB1 by Cameron and Trivedi 1998). In the statistical literature, the reparameterization

$$\mathsf{Var}(y_i|x_i) \quad = \quad (1+\alpha) \cdot \mu_i = \text{dispersion} \cdot \mu_i \tag{5.4}$$

of the NB1 model is often called a quasi-Poisson model with dispersion parameter.

The package **AER** provides the function `dispersiontest()` for testing equidispersion against alternative (5.3). By default, the parameterization (5.4) is used. Alternatively, if the argument `trafo` is specified, the test is formulated in terms of the parameter α. Tests against the quasi-Poisson formulation and the NB2 alternative are therefore available using

```
R> dispersiontest(rd_pois)

        Overdispersion test

data:  rd_pois
z = 2.412, p-value = 0.007941
alternative hypothesis: true dispersion is greater than 1
sample estimates:
dispersion
     6.566
```

and

```
R> dispersiontest(rd_pois, trafo = 2)

        Overdispersion test

data:  rd_pois
z = 2.938, p-value = 0.001651
alternative hypothesis: true alpha is greater than 0
```

```
sample estimates:
alpha
1.316
```

Both suggest that the Poisson model for the trips data is not well specified in that there appears to be a substantial amount of overdispersion.

One possible remedy is to consider a more flexible distribution that does not impose equality of mean and variance. The most widely used distribution in this context is the negative binomial. It may be considered a mixture distribution arising from a Poisson distribution with random scale, the latter following a gamma distribution. Its probability mass function is

$$f(y; \mu, \theta) = \frac{\Gamma(\theta + y)}{\Gamma(\theta) y!} \frac{\mu^y \theta^\theta}{(\mu + \theta)^{y+\theta}}, \quad y = 0, 1, 2, \ldots, \mu > 0, \theta > 0.$$

The variance of the negative binomial distribution is given by

$$\mathsf{Var}(y; \mu, \theta) = \mu + \frac{1}{\theta} \mu^2,$$

which is of the form (5.3) with $h(\mu) = \mu^2$ and $\alpha = 1/\theta$.

If θ is known, this distribution is of the general form (5.1). The Poisson distribution with parameter μ arises for $\theta \to \infty$. The geometric distribution is the special case where $\theta = 1$.

In R, tools for negative binomial regression are provided by the **MASS** package (Venables and Ripley 2002). Specifically, for estimating negative binomial GLMs with known θ, the function `negative.binomial()` can be used; e.g., for a geometric regression via `family = negative.binomial(theta = 1)`. For unknown θ, the function `glm.nb()` is available. Thus

```
R> library("MASS")
R> rd_nb <- glm.nb(trips ~ ., data = RecreationDemand)
R> coeftest(rd_nb)
```

```
z test of coefficients:
```

	Estimate	Std. Error	z value	Pr(>|z|)
(Intercept)	-1.12194	0.21430	-5.24	1.6e-07
quality	0.72200	0.04012	18.00	< 2e-16
skiyes	0.61214	0.15030	4.07	4.6e-05
income	-0.02606	0.04245	-0.61	0.539
userfeeyes	0.66917	0.35302	1.90	0.058
costC	0.04801	0.00918	5.23	1.7e-07
costS	-0.09269	0.00665	-13.93	< 2e-16
costH	0.03884	0.00775	5.01	5.4e-07

```
R> logLik(rd_nb)
```

'log Lik.' -825.6 (df=9)

provides a new model for the trips data. The shape parameter of the fitted negative binomial distribution is $\hat{\theta} = 0.7293$, suggesting a considerable amount of overdispersion and confirming the results from the test for overdispersion. We note that the negative binomial model represents a substantial improvement in terms of likelihood.

Robust standard errors

Instead of switching to a more general family of distributions, it is also possible to use the Poisson estimates along with a set of standard errors estimated under less restrictive assumptions, provided the mean is correctly specified. The package **sandwich** implements sandwich variance estimators for GLMs via the sandwich() function. In microeconometric applications, the resulting estimates are often called Huber-White standard errors. For the problem at hand, Poisson and Huber-White errors are rather different:

```
R> round(sqrt(rbind(diag(vcov(rd_pois)),
+     diag(sandwich(rd_pois)))), digits = 3)
```

	(Intercept)	quality	skiyes	income	userfeeyes	costC	costS
[1,]	0.094	0.017	0.057	0.02	0.079	0.003	0.002
[2,]	0.432	0.049	0.194	0.05	0.247	0.015	0.012

	costH
[1,]	0.003
[2,]	0.009

This again indicates that the simple Poisson model is inadequate. Regression output utilizing robust standard errors is available using

```
R> coeftest(rd_pois, vcov = sandwich)
```

z test of coefficients:

	Estimate	Std. Error	z value	Pr(>|z|)
(Intercept)	0.26499	0.43248	0.61	0.54006
quality	0.47173	0.04885	9.66	< 2e-16
skiyes	0.41821	0.19387	2.16	0.03099
income	-0.11132	0.05031	-2.21	0.02691
userfeeyes	0.89817	0.24691	3.64	0.00028
costC	-0.00343	0.01470	-0.23	0.81549
costS	-0.04254	0.01173	-3.62	0.00029
costH	0.03613	0.00939	3.85	0.00012

It should be noted that the sandwich estimator is based on first and second derivatives of the likelihood: the outer product of gradients (OPG) forms the

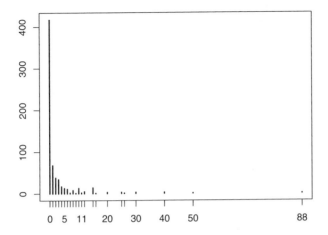

Fig. 5.3. Empirical distribution of trips.

"meat" and the second derivatives the "bread" of the sandwich (see Zeileis 2006b, for further details on the implementation). Both are asymptotically equivalent under correct specification, and the default estimator of the covariance (computed by the vcov() method) is based on the "bread". However, it is also possible to employ the "meat" alone. This is usually called the OPG estimator in econometrics. It is nowadays considered to be inferior to competing estimators, but as it is often found in the older literature, it might be useful to have it available for comparison. OPG standard errors for our model are given by

```
R> round(sqrt(diag(vcovOPG(rd_pois))), 3)
```

(Intercept)	quality	skiyes	income	userfeeyes
0.025	0.007	0.020	0.010	0.033
costC	costS	costH		
0.001	0.000	0.001		

Zero-inflated Poisson and negative binomial models

A typical problem with count data regressions arising in microeconometric applications is that the number of zeros is often much larger than a Poisson or negative binomial regression permit. Figure 5.3 indicates that the RecreationDemand data contain a large number of zeros; in fact, no less than 63.28% of all respondents report no trips to Lake Somerville.

However, the Poisson regression fitted above only provides 41.96% of zero observations, again suggesting that this model is not satisfactory. A closer look at observed and expected counts reveals

```
R> rbind(obs = table(RecreationDemand$trips)[1:10], exp = round(
+     sapply(0:9, function(x) sum(dpois(x, fitted(rd_pois))))))
```

```
      0    1   2   3   4   5   6   7  8 9
obs 417   68  38  34  17  13  11   2  8 1
exp 277  146  68  41  30  23  17  13 10 7
```

This again underlines that there are problems with the previously considered specification. In fact, the variance-to-mean ratio for the `trips` variable equals 17.64, a value usually too large to be accommodated by covariates. Alternative models are needed.

One such model is the zero-inflated Poisson (ZIP) model (Lambert 1992), which suggests a mixture specification with a Poisson count component and an additional point mass at zero. With $I_A(y)$ denoting the indicator function, the basic idea is

$$f_{\text{zeroinfl}}(y) = p_i \cdot I_{\{0\}}(y) \; + \; (1 - p_i) \cdot f_{\text{count}}(y; \mu_i),$$

where now μ_i and p_i are modeled as functions of the available covariates. Employing the canonical link for the count part gives $\log(\mu_i) = x_i^\top \beta$, while for the binary part $g(p_i) = z_i^\top \gamma$ for some quantile function g, with the canonical link given by the logistic distribution and the probit link as a popular alternative. Note that the two sets of regressors x_i and z_i need not be identical.

The package **pscl** provides a function `zeroinfl()` for fitting this family of models (Zeileis, Kleiber, and Jackman 2008), allowing Poisson, geometric, and negative binomial distributions for the count component $f_{\text{count}}(y)$.

Following Cameron and Trivedi (1998), we consider a regression of `trips` on all further variables for the count part (using a negative binomial distribution) and model the inflation part as a function of `quality` and `income`:

```
R> rd_zinb <- zeroinfl(trips ~ . | quality + income,
+     data = RecreationDemand, dist = "negbin")
```

This specifies the linear predictor in the count part as employing all available regressors, while the inflation part, separated by |, only uses `quality` and `income`. The inflation part by default uses the logit model, but all other links available in `glm()` may also be used upon setting the argument `link` appropriately. For the `RecreationDemand` data, a ZINB model yields

```
R> summary(rd_zinb)
```

```
Call:
zeroinfl(formula = trips ~ . | quality + income,
   data = RecreationDemand, dist = "negbin")
```

```
Count model coefficients (negbin with log link):
             Estimate Std. Error z value Pr(>|z|)
(Intercept)   1.09663    0.25668    4.27 1.9e-05
quality       0.16891    0.05303    3.19 0.00145
skiyes        0.50069    0.13449    3.72 0.00020
income       -0.06927    0.04380   -1.58 0.11378
userfeeyes    0.54279    0.28280    1.92 0.05494
costC         0.04044    0.01452    2.79 0.00534
costS        -0.06621    0.00775   -8.55  < 2e-16
costH         0.02060    0.01023    2.01 0.04415
Log(theta)    0.19017    0.11299    1.68 0.09235

Zero-inflation model coefficients (binomial with logit link):
            Estimate Std. Error z value Pr(>|z|)
(Intercept)    5.743      1.556    3.69 0.00022
quality       -8.307      3.682   -2.26 0.02404
income        -0.258      0.282   -0.92 0.35950

Theta = 1.209
Number of iterations in BFGS optimization: 26
Log-likelihood: -722 on 12 Df
```

showing a clearly improved log-likelihood compared with the plain negative binomial model. The expected counts are given by

```
R> round(colSums(predict(rd_zinb, type = "prob")[,1:10]))
```

```
  0    1    2    3    4    5    6    7   8   9
433   47   35   27   20   16   12   10   8   7
```

which shows that both the zeros and the remaining counts are captured much better than in the previous models. Note that the `predict()` method for `type = "prob"` returns a matrix that provides a vector of expected probabilities for each observation. By taking column sums, the expected counts can be computed, here for $0, \ldots, 9$.

Hurdle models

Another model that is able to capture excessively large (or small) numbers of zeros is the hurdle model (Mullahy 1986). In economics, it is more widely used than the zero-inflation model presented above. The hurdle model consists of two parts (hence it is also called a "two-part model"):

- A binary part (given by a count distribution right-censored at $y = 1$): Is y_i equal to zero or positive? "Is the hurdle crossed?"
- A count part (given by a count distribution left-truncated at $y = 1$): If $y_i > 0$, how large is y_i?

This results in a model with

$$f_{\text{hurdle}}(y; x, z, \beta, \gamma)$$
$$= \begin{cases} f_{\text{zero}}(0; z, \gamma), & \text{if } y = 0, \\ \{1 - f_{\text{zero}}(0; z, \gamma)\} \cdot f_{\text{count}}(y; x, \beta)/\{1 - f_{\text{count}}(0; x, \beta)\}, & \text{if } y > 0. \end{cases}$$

The hurdle model is somewhat easier to fit than the zero-inflation model because the resulting likelihood decomposes into two parts that may be maximized separately.

The package **pscl** also provides a function `hurdle()`. Again following Cameron and Trivedi (1998), we consider a regression of `trips` on all further variables for the count part and model the inflation part as a function of only `quality` and `income`. The available distributions in `hurdle()` are again Poisson, negative binomial, and geometric.

We warn readers that there exist several parameterizations of hurdle models that differ with respect to the hurdle mechanism. Here it is possible to specify either a count distribution right-censored at one or a Bernoulli distribution distinguishing between zeros and non-zeros (which is equivalent to the right-censored geometric distribution). The first variant appears to be more common in econometrics; however, the Bernoulli specification is more intuitively interpretable as a hurdle (and hence used by default in `hurdle()`). Here, we employ the latter, which is equivalent to the geometric distribution used by Cameron and Trivedi (1998).

```
R> rd_hurdle <- hurdle(trips ~ . | quality + income,
+     data = RecreationDemand, dist = "negbin")
R> summary(rd_hurdle)

Call:
hurdle(formula = trips ~ . | quality + income,
   data = RecreationDemand, dist = "negbin")


Count model coefficients (truncated negbin with log link):
            Estimate Std. Error z value Pr(>|z|)
(Intercept)   0.8419     0.3828    2.20   0.0278
quality       0.1717     0.0723    2.37   0.0176
skiyes        0.6224     0.1901    3.27   0.0011
income       -0.0571     0.0645   -0.88   0.3763
userfeeyes    0.5763     0.3851    1.50   0.1345
costC         0.0571     0.0217    2.63   0.0085
costS        -0.0775     0.0115   -6.71  1.9e-11
costH         0.0124     0.0149    0.83   0.4064
Log(theta)   -0.5303     0.2611   -2.03   0.0423
```

```
Zero hurdle model coefficients (binomial with logit link):
            Estimate Std. Error z value Pr(>|z|)
(Intercept)  -2.7663     0.3623   -7.64  2.3e-14
quality       1.5029     0.1003   14.98  < 2e-16
income       -0.0447     0.0785   -0.57    0.57
```

```
Theta: count = 0.588
Number of iterations in BFGS optimization: 18
Log-likelihood: -765 on 12 Df
```

Again, we consider the expected counts:

```
R> round(colSums(predict(rd_hurdle, type = "prob")[,1:10]))
```

```
  0   1   2   3   4   5   6   7   8   9
417  74  42  27  19  14  10   8   6   5
```

Like the ZINB model, the negative binomial hurdle model results in a considerable improvement over the initial Poisson specification.

Further details on fitting, assessing and comparing regression models for count data can be found in Zeileis *et al.* (2008).

5.4 Censored Dependent Variables

Censored dependent variables have been used in econometrics ever since Tobin (1958) modeled the demand for durable goods by means of a censored latent variable. Specifically, the tobit model posits a Gaussian linear model for a latent variable, say y^0. y^0 is only observed if positive; otherwise zero is reported:

$$y_i^0 = x_i^\top \beta + \varepsilon_i, \qquad \varepsilon_i | x_i \sim \mathcal{N}(0, \sigma^2) \text{ i.i.d.,}$$
$$y_i = \begin{cases} y_i^0, & y_i^0 > 0, \\ 0, & y_i^0 \le 0. \end{cases}$$

The log-likelihood of this model is thus given by

$$\ell(\beta, \sigma^2) = \sum_{y_i > 0} \left(\log \phi\{(y_i - x_i^\top \beta)/\sigma\} - \log \sigma \right) + \sum_{y_i = 0} \log \Phi(x_i^\top \beta/\sigma).$$

To statisticians, especially biostatisticians, this is a special case of a censored regression model. An R package for fitting such models has long been available, which is the **survival** package accompanying Therneau and Grambsch (2000). (In fact, **survival** even contains Tobin's original data; readers may want to explore data("tobin") there.) However, many aspects of that package will not look familiar to econometricians, and thus our package **AER** comes with a convenience function tobit() interfacing **survival**'s survreg() function.

We illustrate the mechanics using a famous—or perhaps infamous—data set taken from Fair (1978). It is available as `Affairs` in **AER** and provides the results of a survey on extramarital affairs conducted by the US magazine *Psychology Today* in 1969. The dependent variable is `affairs`, the number of extramarital affairs during the past year, and the available regressors include `age`, `yearsmarried`, `children` (a factor indicating the presence of children), `occupation` (a numeric score coding occupation), and `rating` (a numeric variable coding the self-rating of the marriage; the scale is 1 through 5, with 5 indicating "very happy").

We note that the dependent variable is a count and is thus perhaps best analyzed along different lines; however, for historical reasons and for comparison with sources such as Greene (2003), we follow the classical approach. (Readers are asked to explore count data models in an exercise.)

`tobit()` has a formula interface like `lm()`. For the classical tobit model with left-censoring at zero, only this formula and a data set are required:

```
R> data("Affairs")
R> aff_tob <- tobit(affairs ~ age + yearsmarried +
+    religiousness + occupation + rating, data = Affairs)
R> summary(aff_tob)

Call:
tobit(formula = affairs ~ age + yearsmarried +
  religiousness + occupation + rating, data = Affairs)

Observations:
         Total  Left-censored    Uncensored  Right-censored
           601            451           150               0

Coefficients:
              Estimate Std. Error z value Pr(>|z|)
(Intercept)     8.1742     2.7414    2.98   0.0029
age            -0.1793     0.0791   -2.27   0.0234
yearsmarried    0.5541     0.1345    4.12  3.8e-05
religiousness  -1.6862     0.4038   -4.18  3.0e-05
occupation      0.3261     0.2544    1.28   0.2000
rating         -2.2850     0.4078   -5.60  2.1e-08
Log(scale)      2.1099     0.0671   31.44  < 2e-16

Scale: 8.25

Gaussian distribution
Number of Newton-Raphson Iterations: 4
Log-likelihood: -706 on 7 Df
Wald-statistic: 67.7 on 5 Df, p-value: 3.1e-13
```

The output comprises the usual regression output along with the value of the log-likelihood and a Wald statistic paralleling the familiar regression F statistic. For convenience, a tabulation of censored and uncensored observations is also included. The results indicate that yearsmarried and rating are the main "risk factors".

To further illustrate the arguments to tobit(), we refit the model by introducing additional censoring from the right:

```
R> aff_tob2 <- update(aff_tob, right = 4)
R> summary(aff_tob2)

Call:
tobit(formula = affairs ~ age + yearsmarried +
  religiousness + occupation + rating, right = 4,
  data = Affairs)

Observations:
          Total  Left-censored     Uncensored Right-censored
            601            451             70             80

Coefficients:
              Estimate Std. Error z value Pr(>|z|)
(Intercept)     7.9010     2.8039    2.82  0.00483
age            -0.1776     0.0799   -2.22  0.02624
yearsmarried    0.5323     0.1412    3.77  0.00016
religiousness  -1.6163     0.4244   -3.81  0.00014
occupation      0.3242     0.2539    1.28  0.20162
rating         -2.2070     0.4498   -4.91  9.3e-07
Log(scale)      2.0723     0.1104   18.77  < 2e-16

Scale: 7.94

Gaussian distribution
Number of Newton-Raphson Iterations: 4
Log-likelihood: -500 on 7 Df
Wald-statistic: 42.6 on 5 Df, p-value: 4.5e-08
```

The standard errors are now somewhat larger, reflecting the fact that heavier censoring leads to a loss of information. tobit() also permits, via the argument dist, alternative distributions of the latent variable, including the logistic and Weibull distributions.

Among the methods for objects of class "tobit", we briefly consider a Wald-type test:

```
R> linear.hypothesis(aff_tob, c("age = 0", "occupation = 0"),
+     vcov = sandwich)
```

Linear hypothesis test

Hypothesis:
age = 0
occupation = 0

Model 1: affairs ~ age + yearsmarried + religiousness +
   occupation + rating
Model 2: restricted model

Note: Coefficient covariance matrix supplied.

```
  Res.Df  Df Chisq Pr(>Chisq)
1    594
2    596  -2  4.91       0.086
```

Thus, the regressors age and occupation are jointly weakly significant. For illustration, we use a sandwich covariance estimate, although it should be borne in mind that, as in the binary and unlike the Poisson case, in this model, misspecification of the variance typically also means misspecification of the mean (see again Freedman 2006, for further discussion).

5.5 Extensions

The number of models used in microeconometrics has grown considerably over the last two decades. Due to space constraints, we can only afford to briefly discuss a small selection. We consider a semiparametric version of the binary response model as well as multinomial and ordered logit models.

Table 5.2 provides a list of further relevant packages.

A semiparametric binary response model

Recall that the log-likelihood of the binary response model is

$$\ell(\beta) = \sum_{i=1}^{n} \left\{ y_i \log F(x_i^\top \beta) + (1 - y_i) \log\{1 - F(x_i^\top \beta)\} \right\},$$

where F is the CDF of the logistic or the Gaussian distribution in the logit or probit case, respectively. The Klein and Spady (1993) approach estimates F via kernel methods, and thus it may be considered a semiparametric maximum likelihood estimator. In another terminology, it is a semiparametric single-index model. We refer to Li and Racine (2007) for a recent exposition.

In R, the Klein and Spady estimator is available in the package **np** (Hayfield and Racine 2008), the package accompanying Li and Racine (2007). Since the required functions from that package currently do not accept factors as dependent variables, we preprocess the SwissLabor data via

Table 5.2. Further packages for microeconometrics.

Package	Description
gam	Generalized additive models (Hastie 2006)
lme4	Nonlinear random-effects models: counts, binary dependent variables, etc. (Bates 2008)
mgcv	Generalized additive (mixed) models (Wood 2006)
micEcon	Demand systems, cost and production functions (Henningsen 2008)
mlogit	Multinomial logit models with choice-specific variables (Croissant 2008)
robustbase	Robust/resistant regression for GLMs (Maechler, Rousseeuw, Croux, Todorov, Ruckstuhl, and Salibian-Barrera 2007)
sampleSelection	Selection models: generalized tobit, heckit (Toomet and Henningsen 2008)

```
R> SwissLabor$partnum <- as.numeric(SwissLabor$participation) - 1
```

which creates a dummy variable `partnum` within `SwissLabor` that codes nonparticipation and participation as 0 and 1, respectively. Fitting itself requires first computing a bandwidth object using the function `npindexbw()`, as in

```
R> library("np")
R> swiss_bw <- npindexbw(partnum ~ income + age + education +
+    youngkids + oldkids + foreign + I(age^2), data = SwissLabor,
+    method = "kleinspady", nmulti = 5)
```

A summary of the bandwidths is available via

```
R> summary(swiss_bw)

Single Index Model
Regression data (872 observations, 7 variable(s)):

       income    age education youngkids oldkids foreign
Beta:       1 -2.219   -0.0249    -5.515  0.1797 -0.8268
       I(age^2)
Beta:    0.3427
Bandwidth:  0.383
Optimisation Method:  Nelder-Mead
Regression Type: Local-Constant
Bandwidth Selection Method: Klein and Spady
```

```
Formula: partnum ~ income + age + education + youngkids +
  oldkids + foreign + I(age^2)
Objective Function Value: 0.5934 (achieved on multistart 3)

Continuous Kernel Type: Second-Order Gaussian
No. Continuous Explanatory Vars.: 1
```

Finally, the Klein and Spady estimate is given by passing the bandwidth object swiss_bw to npindex():

```
R> swiss_ks <- npindex(bws = swiss_bw, gradients = TRUE)
R> summary(swiss_ks)

Single Index Model
Regression Data: 872 training points, in 7 variable(s)

        income     age education youngkids oldkids foreign
Beta:        1 -2.219   -0.0249    -5.515  0.1797 -0.8268
     I(age^2)
Beta:   0.3427
Bandwidth: 0.383
Kernel Regression Estimator: Local-Constant

Confusion Matrix
      Predicted
Actual   0   1
     0 345 126
     1 137 264

Overall Correct Classification Ratio:  0.6984
Correct Classification Ratio By Outcome:
     0      1
0.7325 0.6584

McFadden-Puig-Kerschner performance measure from
prediction-realization tables:  0.6528

Continuous Kernel Type: Second-Order Gaussian
No. Continuous Explanatory Vars.: 1
```

The resulting confusion matrix may be compared with the confusion matrix of the original probit model (see Section 5.2),

```
R> table(Actual = SwissLabor$participation, Predicted =
+     round(predict(swiss_probit, type = "response")))
```

```
        Predicted
Actual    0    1
   no   337  134
   yes  146  255
```

showing that the semiparametric model has slightly better (in-sample) performance.

When applying semiparametric procedures such as the Klein and Spady method, one should be aware that these are rather time-consuming (despite the optimized and compiled C code underlying the **np** package). In fact, the model above takes more time than all other examples together when compiling this book on the authors' machines.

Multinomial responses

For illustrating the most basic version of the multinomial logit model, a model with only individual-specific covariates, we consider the BankWages data taken from Heij, de Boer, Franses, Kloek, and van Dijk (2004). It contains, for employees of a US bank, an ordered factor job with levels "custodial", "admin" (for administration), and "manage" (for management), to be modeled as a function of education (in years) and a factor minority indicating minority status. There also exists a factor gender, but since there are no women in the category "custodial", only a subset of the data corresponding to males is used for parametric modeling below.

To obtain a first overview of how job depends on education, a table of conditional proportions can be generated via

```
R> data("BankWages")
R> edcat <- factor(BankWages$education)
R> levels(edcat)[3:10] <- rep(c("14-15", "16-18", "19-21"),
+    c(2, 3, 3))
R> tab <- xtabs(~ edcat + job, data = BankWages)
R> prop.table(tab, 1)
```

```
        job
edcat    custodial     admin    manage
   8      0.245283  0.754717  0.000000
  12      0.068421  0.926316  0.005263
 14-15    0.008197  0.959016  0.032787
 16-18    0.000000  0.367089  0.632911
 19-21    0.000000  0.033333  0.966667
```

where education has been transformed into a categorical variable with some of the sparser levels merged. This table can also be visualized in a spine plot via

```
R> plot(job ~ edcat, data = BankWages, off = 0)
```

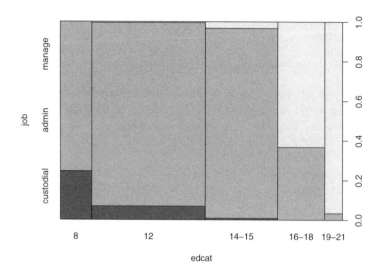

Fig. 5.4. Relationship between job category and education.

or equivalently via `spineplot(tab, off = 0)`. The result in Figure 5.4 indicates that the proportion of `"custodial"` employees quickly decreases with education and that, at higher levels of education, a larger proportion of individuals is employed in the management category.

Multinomial logit models permit us to quantify this observation. They can be fitted utilizing the `multinom()` function from the package **nnet** (for "neural networks"), a package from the **VR** bundle accompanying Venables and Ripley (2002). Note that the function is only superficially related to neural networks in that the algorithm employed is the same as that for single hidden-layer neural networks (as provided by `nnet()`).

The main arguments to `multinom()` are again `formula` and `data`, and thus a multinomial logit model is fitted via

```
R> library("nnet")
R> bank_mnl <- multinom(job ~ education + minority,
+     data = BankWages, subset = gender == "male", trace = FALSE)
```

Instead of providing the full `summary()` of the fit, we just give the more compact

```
R> coeftest(bank_mnl)
```

```
z test of coefficients:
```

	Estimate Std.	Error	z value	Pr(>|z|)
admin:(Intercept)	-4.761	1.173	-4.06	4.9e-05
admin:education	0.553	0.099	5.59	2.3e-08
admin:minorityyes	-0.427	0.503	-0.85	0.3957
manage:(Intercept)	-30.775	4.479	-6.87	6.4e-12
manage:education	2.187	0.295	7.42	1.2e-13
manage:minorityyes	-2.536	0.934	-2.71	0.0066

This confirms that the proportions of "admin" and "manage" job categories (as compared with the reference category, here "custodial") increase with education and decrease for minority. Both effects seem to be stronger for the "manage" category.

We add that, in contrast to multinom(), the recent package **mlogit** (Croissant 2008) also fits multinomial logit models containing "choice-specific" (i.e., outcome-specific) attributes.

Ordinal responses

The dependent variable job in the preceding example can be considered an ordered response, with "custodial" < "admin" < "manage". This suggests that an ordered logit or probit regression may be worth exploring; here we consider the former. In the statistical literature, this is often called proportional odds logistic regression; hence the name polr() for the fitting function from the **MASS** package (which, despite its name, can also fit ordered probit models upon setting method="probit"). Here, this yields

```
R> library("MASS")
R> bank_polr <- polr(job ~ education + minority,
+    data = BankWages, subset = gender == "male", Hess = TRUE)
R> coeftest(bank_polr)
```

```
z test of coefficients:
```

	Estimate Std.	Error	z value	Pr(>|z|)
education	0.8700	0.0931	9.35	< 2e-16
minorityyes	-1.0564	0.4120	-2.56	0.010
custodial|admin	7.9514	1.0769	7.38	1.5e-13
admin|manage	14.1721	0.0941	150.65	< 2e-16

using again the more concise output of coeftest() rather than summary(). The ordered logit model just estimates different intercepts for the different job categories but a common set of regression coefficients. The results are similar to those for the multinomial model, but the different education and minority effects for the different job categories are, of course, lost. This appears to deteriorate the model fit as the AIC increases:

```
R> AIC(bank_mnl)
```

```
[1] 249.5
```

```
R> AIC(bank_polr)
```

```
[1] 268.6
```

5.6 Exercises

1. For the SwissLabor data, plotting participation versus education (see Figure 5.1) suggests a nonlinear effect of education. Fit a model utilizing education squared in addition to the regressors considered in Section 5.2. Does the new model result in an improvement?

2. The PSID1976 data originating from Mroz (1987) are used in many econometrics texts, including Greene (2003) and Wooldridge (2002). Following Greene (2003, p. 681):

   (a) Fit a probit model for labor force participation using the regressors age, age squared, family income, education, and a factor indicating the presence of children. (The factor needs to be constructed from the available information.)

   (b) Reestimate the model assuming that different equations apply to women with and without children.

   (c) Perform a likelihood ratio test to check whether the more general model is really needed.

3. Analyze the DoctorVisits data, taken from Cameron and Trivedi (1998), using a Poisson regression for the number of visits. Is the Possion model satisfactory? If not, where are the problems and what could be done about them?

4. As mentioned above, the Affairs data are perhaps better analyzed utilizing models for count data rather than a tobit model as we did here. Explore a Poisson regression and some of its variants, and be sure to check whether the models accommodate the many zeros present in these data.

5. Using the PSID1976 data, run a tobit regression of hours worked on non-wife income (to be constructed from the available information), age, experience, experience squared, education, and the numbers of younger and older children.

6

Time Series

Time series arise in many fields of economics, especially in macroeconomics and financial economics. Here, we denote a time series (univariate or multivariate) as y_t, $t = 1, \ldots, n$. This chapter first provides a brief overview of R's time series classes and "naive" methods such as the classical decomposition into a trend, a seasonal component, and a remainder term, as well as exponential smoothing and related techniques. It then moves on to autoregressive moving average (ARMA) models and extensions. We discuss classical Box-Jenkins style analysis based on the autocorrelation and partial autocorrelation functions (ACF and PACF) as well as model selection via information criteria.

Many time series in economics are nonstationary. Nonstationarity often comes in one of two forms: the time series can be reduced to stationarity by differencing or detrending, or it contains structural breaks and is therefore only piecewise stationary. The third section therefore shows how to perform the standard unit-root and stationarity tests as well as cointegration tests. The fourth section discusses the analysis of structural change, where R offers a particularly rich set of tools for testing as well as dating breaks. The final section briefly discusses structural time series models and volatility models.

Due to space constraints, we confine ourselves to time domain methods. However, all the standard tools for analysis in the frequency domain, notably estimates of the spectral density by several techniques, are available as well. In fact, some of these methods have already been used, albeit implicitly, in connection with HAC covariance estimation in Chapter 4.

6.1 Infrastructure and "Naive" Methods

Classes for time series data

In the previous chapters, we already worked with different data structures that can hold rectangular data matrices, most notably "data.frame" for

C. Kleiber, A. Zeileis, *Applied Econometrics with R*,
DOI: 10.1007/978-0-387-77318-6_6, © Springer Science+Business Media, LLC 2008

cross-sectional data. Dealing with time series data poses slightly different challenges. While we also need a rectangular, typically numeric, data matrix, in addition, some way of storing the associated time points of the series is required. R offers several classes for holding such data. Here, we discuss the two most important (closely related) classes, "ts" and "zoo".

R ships with the basic class "ts" for representing time series data; it is aimed at regular series, in particular at annual, quarterly, and monthly data. Objects of class "ts" are either a numeric vector (for univariate series) or a numeric matrix (for multivariate series) containing the data, along with a "tsp" attribute reflecting the time series properties. This is a vector of length three containing the start and end times (in time units) and the frequency. Time series objects of class "ts" can easily be created with the function ts() by supplying the data (a numeric vector or matrix), along with the arguments start, end, and frequency. Methods for standard generic functions such as plot(), lines(), str(), and summary() are provided as well as various time-series-specific methods, such as lag() or diff(). As an example, we load and plot the univariate time series UKNonDurables, containing the quarterly consumption of non-durables in the United Kingdom (taken from Franses 1998).

```
R> data("UKNonDurables")
R> plot(UKNonDurables)
```

The resulting time series plot is shown in the left panel of Figure 6.1. The time series properties

```
R> tsp(UKNonDurables)
```

```
[1] 1955.00 1988.75    4.00
```

reveal that this is a quarterly series starting in 1955(1) and ending in 1988(4). If the series of all time points is needed, it can be extracted via time(); e.g., time(UKNonDurables). Subsets can be chosen using the function window(); e.g.,

```
R> window(UKNonDurables, end = c(1956, 4))
```

```
      Qtr1  Qtr2  Qtr3  Qtr4
1955 24030 25620 26209 27167
1956 24620 25972 26285 27659
```

Single observations can be extracted by setting start and end to the same value.

The "ts" class is well suited for annual, quarterly, and monthly time series. However, it has two drawbacks that make it difficult to use in some applications: (1) it can only deal with numeric time stamps (and not with more general date/time classes); (2) internal missing values cannot be omitted (because then the start/end/frequency triple is no longer sufficient for

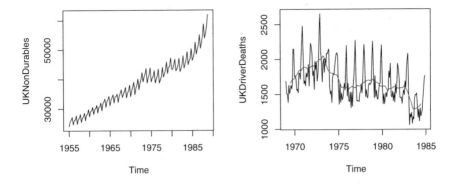

Fig. 6.1. Quarterly time series of consumption of non-durables in the United Kingdom (left) and monthly number of car drivers killed or seriously injured in the United Kingdom (right, with filtered version).

reconstructing all time stamps). Both are major nuisances when working with irregular series (e.g., with many financial time series). Consequently, various implementations for irregular time series have emerged in contributed R packages, the most flexible of which is "zoo", provided by the **zoo**[1] package (Zeileis and Grothendieck 2005). It can have time stamps of arbitrary type and is designed to be as similar as possible to "ts". Specifically, the series are essentially also numeric vectors or matrices but with an "index" attribute containing the full vector of indexes (or time stamps) rather than only the "tsp" attribute with start/end/frequency. Therefore, "zoo" series can be seen as a generalization of "ts" series. Most methods that work for "ts" objects also work for "zoo" objects, some have extended functionality, and some new ones are provided. Regular series can be coerced back and forth between the classes without loss of information using the functions as.zoo() and as.ts(). Hence, it does not make much difference which of these two classes is used for annual, quarterly, or monthly data—whereas "zoo" is much more convenient for daily data (e.g., coupled with an index of class "Date") or intraday data (e.g., with "POSIXct" or "chron" time stamps). See Zeileis and Grothendieck (2005) for further details on "zoo" series and Grothendieck and Petzoldt (2004) for more on date/time classes in R.

Throughout this book, we mainly rely on the "ts" class; only in very few illustrations where additional flexibility is required do we switch to "zoo".

---

[1] **zoo** stands for Z's ordered observations, named after the author who started the development of the package.

(Linear) filtering

One of the most basic tools for transforming time series (e.g., for eliminating seasonality) is linear filtering (see Brockwell and Davis 1991, 1996). An important class of linear filters are finite moving averages, transformations that replace the raw data y_t by a weighted sum

$$\hat{y}_t = \sum_{j=-r}^{s} a_j y_{t+j}, \quad t = r+1, \ldots, n-s.$$

If r equals s, the filter is called symmetric. In R, the function `filter()` permits the use of fairly general filters; its argument `filter` takes a vector containing the coefficients a_j. Apart from moving averages (default, see above), `filter()` can also apply recursive linear filters, another important class of filters.

As an example, we consider the monthly time series `UKDriverDeaths` containing the well-known data from Harvey and Durbin (1986) on car drivers killed or seriously injured in the United Kingdom from 1969(1) through 1984(12). These are also known as the "seatbelt data", as they were used by Harvey and Durbin (1986) for evaluating the effectiveness of compulsory wearing of seatbelts introduced on 1983-01-31. The following code loads and plots the series along with a filtered version utilizing the simple symmetric moving average of length 13 with coefficients $(1/24, 1/12, \ldots, 1/12, 1/24)^\top$.

```
R> data("UKDriverDeaths")
R> plot(UKDriverDeaths)
R> lines(filter(UKDriverDeaths, c(1/2, rep(1, 11), 1/2)/12),
+    col = 2)
```

The resulting plot is depicted in the right panel of Figure 6.1, illustrating that the filter eliminates seasonality. Other classical filters, such as the Henderson or Spencer filters (Brockwell and Davis 1991), can be applied analogously.

Another function that can be used for evaluating linear and nonlinear functions on moving data windows is `rollapply()` (for rolling apply). This can be used for computing running means via `rollapply(UKDriverDeaths, 12, mean)`, yielding a result similar to that for the symmetric filter above, or running standard deviations

```
R> plot(rollapply(UKDriverDeaths, 12, sd))
```

shown in the right panel of Figure 6.2, revealing increased variation around the time of the policy intervention.

As mentioned above, `filter()` also provides autoregressive (recursive) filtering. This can be exploited for simple simulations; e.g., from AR(1) models. The code

```
R> set.seed(1234)
R> x <- filter(rnorm(100), 0.9, method = "recursive")
```

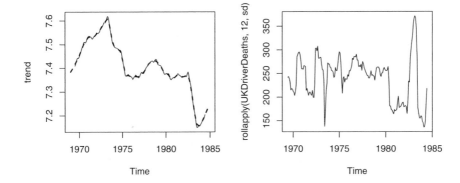

Fig. 6.2. UK driver deaths: trends from two season-trend decompositions (left) and running standard deviations (right).

generates 100 observations from an AR(1) process with parameter 0.9 and standard normal innovations. A more elaborate tool for simulating from general ARIMA models is provided by `arima.sim()`.

Decomposition

Filtering with moving averages can also be used for an additive or multiplicative decomposition into seasonal, trend, and irregular components. The classical approach to this task, implemented in the function `decompose()`, is to take a simple symmetric filter as illustrated above for extracting the trend and derive the seasonal component by averaging the trend-adjusted observations from corresponding periods. A more sophisticated approach that also accommodates time-varying seasonal components is seasonal decomposition via loess smoothing (Cleveland, Cleveland, McRae, and Terpenning 1990). It is available in the function `stl()` and iteratively finds the seasonal and trend components by loess smoothing of the observations in moving data windows of a certain size. Both methods are easily applied in R using (in keeping with the original publication, we employ logarithms)

```
R> dd_dec <- decompose(log(UKDriverDeaths))
R> dd_stl <- stl(log(UKDriverDeaths), s.window = 13)
```

where the resulting objects `dd_dec` and `dd_stl` hold the trend, seasonal, and irregular components in slightly different formats (a list for the "decomposed.ts" and a multivariate time series for the "stl" object). Both classes have plotting methods drawing time series plots of the components with a common time axis. The result of `plot(dd_stl)` is provided in Figure 6.3, and the result of `plot(dd_dec)` looks rather similar. (The bars

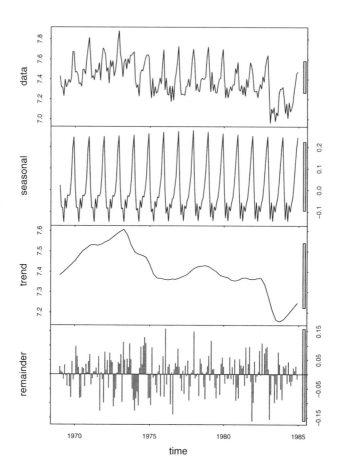

Fig. 6.3. Season-trend decomposition by loess smoothing.

at the right side of Figure 6.3 are of equal heights in user coordinates to ease interpretation.) A graphical comparison of the fitted trend components, obtained via

```
R> plot(dd_dec$trend, ylab = "trend")
R> lines(dd_stl$time.series[,"trend"], lty = 2, lwd = 2)
```

and given in Figure 6.2 (left panel), reveals that both methods provide qualitatively similar results, with stl() yielding a smoother curve. Analogously, the seasonal components dd_stl$time.series[,"seasonal"] and dd_dec$seasonal could be extracted and compared. We note that stl() has artificially smoothed over the structural break due to the introduction of

Holt–Winters filtering

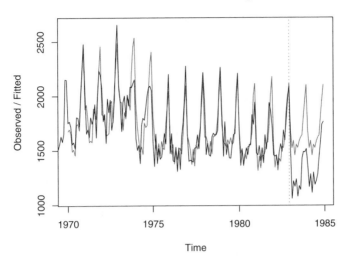

Fig. 6.4. Predictions from Holt-Winters exponential smoothing.

seatbelts, inducing some rather large values in the remainder series. We shall come back to this example in Section 6.4 using different methods.

Exponential smoothing

Another application of linear filtering techniques are classical forecasting methods of the exponential smoothing type, such as simple or double exponential smoothing, employing recursively reweighted lagged observations for predicting future data. A general framework for this is Holt-Winters smoothing (see Meyer 2002, for a brief introduction in R), which comprises these and related exponential smoothing techniques as special cases. The function HoltWinters() implements the general methodology—by default computing a Holt-Winters filter with an additive seasonal component, determining the smoothing parameters by minimizing the squared prediction error on the observed data. To illustrate its use, we separate the UKDriverDeaths series into a historical sample up to 1982(12) (i.e., before the change in legislation) and use Holt-Winters filtering to predict the observations for 1983 and 1984.

```
R> dd_past <- window(UKDriverDeaths, end = c(1982, 12))
R> dd_hw <- HoltWinters(dd_past)
R> dd_pred <- predict(dd_hw, n.ahead = 24)
```

Figure 6.4 compares Holt-Winters predictions with the actually observed series after the policy intervention via

```
R> plot(dd_hw, dd_pred, ylim = range(UKDriverDeaths))
R> lines(UKDriverDeaths)
```

showing that the number of road casualties clearly dropped after the introduction of mandatory wearing of seatbelts.

We conclude by noting that a more sophisticated function for exponential smoothing algorithms, named ets(), is available in the package **forecast**
(Hyndman and Khandakar 2008).

6.2 Classical Model-Based Analysis

The classical approach to parametric modeling and forecasting is to employ
an autoregressive integrated moving average (ARIMA) model for capturing
the dependence structure in a time series (Brockwell and Davis 1991; Box and
Jenkins 1970; Hamilton 1994). To fix the notation, ARIMA(p, d, q) models are
defined by the equation

$$\phi(L)(1 - L)^d y_t = \theta(L)\varepsilon_t, \tag{6.1}$$

where the autoregressive (AR) part is given by the pth-order polynomial
$\phi(L) = 1 - \phi_1 L - \ldots - \phi_p L^p$ in the lag operator L, the moving average (MA)
part is given by the qth-order polynomial $\theta(L) = 1 + \theta_1 L + \ldots + \theta_q L^q$, and d is
the order of differencing. (Note the sign convention for the MA polynomial.)

For ease of reference, Table 6.1 provides a partial list of time series fitting
functions (with StructTS() being discussed in Section 6.5).

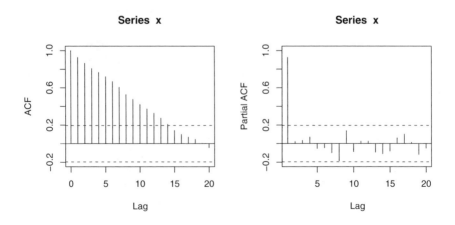

Fig. 6.5. (Partial) autocorrelation function.

Table 6.1. Time series fitting functions in R.

Function	Package	Description
ar()	**stats**	Fits univariate autoregressions via Yule-Walker, OLS, ML, or Burg's method and unrestricted VARs by Yule-Walker, OLS, or Burg's method. Order selection by AIC possible.
arima()	**stats**	Fits univariate ARIMA models, including seasonal (SARIMA) models, models with covariates (ARIMAX), and subset ARIMA models, by unconditional ML or by CSS.
arma()	**tseries**	Fits ARMA models by CSS. Starting values via Hannan-Rissanen. Note: Parameterization of intercept different from arima().
auto.arima()	**forecast**	Order selection via AIC, BIC, or AICC within a user-defined set of models. Fitting is done via arima().
StructTS()	**stats**	Fits structural time series models: local level, local trend, and basic structural model.

Before fitting an ARIMA model to a series, it is helpful to first take an exploratory look at the empirical ACF and PACF. In R, these are available in the functions acf() and pacf(), respectively. For the artificial AR(1) process x from the previous section, they are computed and plotted by

```
R> acf(x)
R> pacf(x)
```

and shown in Figure 6.5. Here, the maximum number of lags defaults to $10 \log_{10}(n)$, but it can be changed by the argument lag.max. Plotting can be suppressed by setting plot = FALSE. This is useful when the (P)ACF is needed for further computations.

The empirical ACF for series x decays only slowly, while the PACF exceeds the individual confidence limits only for lag 1, reflecting clearly how the series was generated. Next we try to recover the true structure by fitting an autoregression to x via the function ar():

```
R> ar(x)

Call:
ar(x = x)

Coefficients:
     1
0.928
```

```
Order selected 1  sigma^2 estimated as  1.29
```

This agrees rather well with the true autocorrelation of 0.9. By default, `ar()` fits AR models up to lag $p = 10 \log_{10}(n)$ and selects the minimum AIC model. This is usually a good starting point for model selection, even though the default estimator is the Yule-Walker estimator, which is considered a preliminary estimator. ML is also available in addition to OLS and Burg estimation; see `?ar` for further details.

For a real-world example, we return to the `UKNonDurables` series, aiming at establishing a model for the (log-transformed) observations up to 1970(4) for predicting the remaining series.

```
R> nd <- window(log(UKNonDurables), end = c(1970, 4))
```

The corresponding time series plot (see Figure 6.1) suggests rather clearly that differencing is appropriate; hence the first row of Figure 6.6 depicts the empirical ACF and PACF of the differenced series. As both exhibit a strong seasonal pattern (already visible in the original series), the second row of Figure 6.6 also shows the empirical ACF and PACF after double differencing (at the seasonal lag 4 and at lag 1) as generated by

```
R> acf(diff(nd), ylim = c(-1, 1))
R> pacf(diff(nd), ylim = c(-1, 1))
R> acf(diff(diff(nd, 4)), ylim = c(-1, 1))
R> pacf(diff(diff(nd, 4)), ylim = c(-1, 1))
```

For this series, a model more general than a simple AR model is needed: `arima()` fits general ARIMA models, including seasonal ARIMA (SARIMA) models and models containing further regressors (utilizing the argument `xreg`), either via ML or minimization of the conditional sum of squares (CSS). The default is to use CSS to obtain starting values and then ML for refinement. As the model space is much more complex than in the AR(p) case, where only the order p has to be chosen, base R does not offer an automatic model selection method for general ARIMA models based on information criteria. Therefore, we use the preliminary results from the exploratory analysis above and R's general tools to set up a model search for an appropriate SARIMA$(p, d, q)(P, D, Q)_4$ model,

$$\Phi(L^4)\phi(L)(1 - L^4)^D(1 - L)^d y_t = \theta(L)\Theta(L^4)\varepsilon_t, \tag{6.2}$$

which amends the standard ARIMA model (6.1) by additional polynomials operating on the seasonal frequency.

The graphical analysis clearly suggests double differencing of the original series ($d = 1$, $D = 1$), some AR and MA effects (we allow $p = 0, 1, 2$ and $q = 0, 1, 2$), and low-order seasonal AR and MA parts (we use $P = 0, 1$ and $Q = 0, 1$), giving a total of 36 parameter combinations to consider. Of course, higher values for p, q, P, and Q could also be assessed. We refrain from doing so

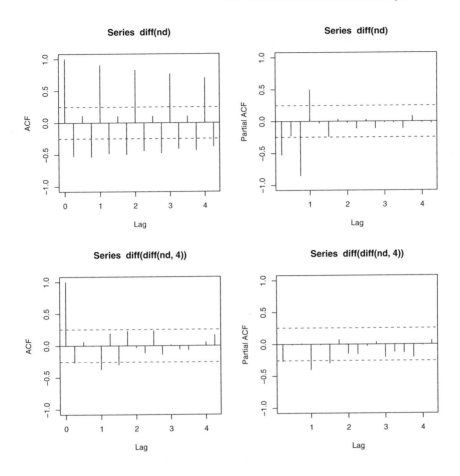

Fig. 6.6. (Partial) autocorrelation functions for UK non-durables data.

to save computation time in this example—however, we encourage readers to pursue this issue for higher-order models as well. We also note that the package **forecast** (Hyndman and Khandakar 2008) contains a function `auto.arima()` that performs a search over a user-defined set of models, just as we do here manually. To choose from the 36 possible models, we set up all parameter combinations via `expand.grid()`, fit each of the associated SARIMA models using `arima()` in a `for()` loop, and store the resulting BIC extracted from the model (obtained via `AIC()` upon setting `k = log(length(nd))`).

```
R> nd_pars <- expand.grid(ar = 0:2, diff = 1, ma = 0:2,
+      sar = 0:1, sdiff = 1, sma = 0:1)
R> nd_aic <- rep(0, nrow(nd_pars))
R> for(i in seq(along = nd_aic)) nd_aic[i] <- AIC(arima(nd,
```

```
+     unlist(nd_pars[i, 1:3]), unlist(nd_pars[i, 4:6])),
+     k = log(length(nd)))
R> nd_pars[which.min(nd_aic),]
```

```
   ar diff ma sar sdiff sma
22 0    1  1  0     1   1
```

These computations reveal that a $SARIMA(0, 1, 1)(0, 1, 1)_4$ model is best in terms of BIC, conforming well with the exploratory analysis. This model is also famously known as the *airline model* due to its application to a series of airline passengers in the classical text by Box and Jenkins (1970). It is refitted to nd via

```
R> nd_arima <- arima(nd, order = c(0,1,1), seasonal = c(0,1,1))
R> nd_arima
```

```
Call:
arima(x = nd, order = c(0, 1, 1), seasonal = c(0, 1, 1))
```

```
Coefficients:
         ma1    sma1
      -0.353  -0.583
s.e.   0.143   0.138
```

```
sigma^2 estimated as 9.65e-05:  log likelihood = 188.14,
aic = -370.27
```

showing that both moving average coefficients are negative and significant. To assess whether this model appropriately captures the dependence structure of the series, tsdiag() produces several diagnostic plots

```
R> tsdiag(nd_arima)
```

shown in Figure 6.7. In the first panel, the standardized residuals are plotted. They do not exhibit any obvious pattern. Their empirical ACF in the second panel shows no (individually) significant autocorrelation at lags > 1. Finally, the p values for the Ljung-Box statistic in the third panel all clearly exceed 5% for all orders, indicating that there is no significant departure from white noise for the residuals.

As there are no obvious deficiencies in our model, it is now used for predicting the remaining 18 years in the sample:

```
R> nd_pred <- predict(nd_arima, n.ahead = 18 * 4)
```

The object nd_pred contains the predictions along with their associated standard errors and can be compared graphically with the observed series via

```
R> plot(log(UKNonDurables))
R> lines(nd_pred$pred, col = 2)
```

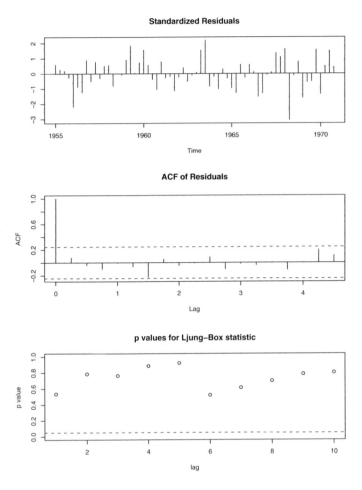

Fig. 6.7. Diagnostics for SARIMA$(0, 1, 1)(0, 1, 1)_4$ model.

Figure 6.8 shows that the general trend is captured reasonably well. However, the model systematically overpredicts for certain parts of the sample in the 1980s.

We conclude by adding that there exist, apart from generic functions such as coef(), logLik(), predict(), print(), and vcov(), which also work on objects of class "Arima" (the class of objects returned by arima()), several further convenience functions for exploring ARMA models and their representations. These are listed in Table 6.2.

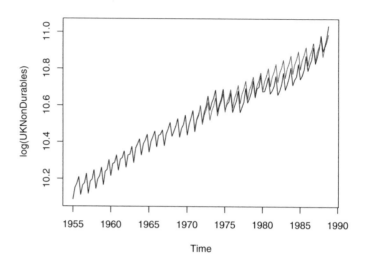

Fig. 6.8. Predictions from SARIMA$(0, 1, 1)(0, 1, 1)_4$ model.

Table 6.2. Convenience functions for ARMA models.

Function	Package	Description
acf2AR()	stats	Computes an AR process exactly fitting a given autocorrelation function.
arima.sim()	stats	Simulation of ARIMA models.
ARMAacf()	stats	Computes theoretical (P)ACF for a given ARMA model.
ARMAtoMA()	stats	Computes MA(∞) representation for a given ARMA model.

6.3 Stationarity, Unit Roots, and Cointegration

Many time series in macroeconomics and finance are nonstationary, the precise form of the nonstationarity having caused a hot debate some 25 years ago. Nelson and Plosser (1982) argued that macroeconomic time series often are more appropriately described by unit-root nonstationarity than by deterministic time trends. We refer to Hamilton (1994) for an overview of the methodology. The Nelson and Plosser data (or rather an extended version ending in 1988) are available in R in the package **tseries** (Trapletti 2008), and we shall use them for an exercise at the end of this chapter.

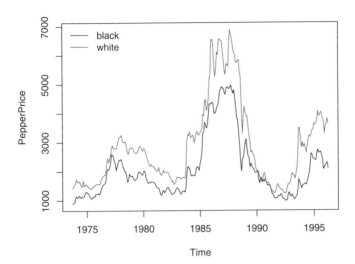

Fig. 6.9. Time series of average monthly European spot prices for black and white pepper (fair average quality) in US dollars per ton.

To illustrate the main methods, we employ a different data set, PepperPrice, containing a bivariate time series of average monthly European spot prices for black and white pepper in US dollars per ton. The data are taken from Franses (1998) and are part of the **AER** package accompanying this book. Figure 6.9 plots both series, and for obvious reasons they are rather similar:

```
R> data("PepperPrice")
R> plot(PepperPrice, plot.type = "single", col = 1:2)
R> legend("topleft", c("black", "white"), bty = "n",
+   col = 1:2, lty = rep(1,2))
```

We begin with an investigation of the time series properties of the individual series, specifically determining their order of integration. There are two ways to proceed: one can either test the null hypothesis of difference stationarity against stationarity (the approach of the classical unit-root tests) or reverse the roles of the alternatives and use a stationarity test such as the KPSS test (Kwiatkowski, Phillips, Schmidt, and Shin 1992).

Unit-root tests

The test most widely used by practitioners, the augmented Dickey-Fuller (ADF) test (Dickey and Fuller 1981), is available in the function adf.test()

from the package **tseries** (Trapletti 2008). This function implements the t test of $H_0 : \varrho = 0$ in the regression

$$\Delta y_t = \alpha + \delta t + \varrho y_{t-1} + \sum_{j=1}^{k} \phi_j \Delta y_{t-j} + \varepsilon_t. \tag{6.3}$$

The number of lags k defaults to $\lfloor (n-1)^{1/3} \rfloor$ but may be changed by the user. For the series corresponding to the price of white pepper, the test yields

```
R> library("tseries")
R> adf.test(log(PepperPrice[, "white"]))

        Augmented Dickey-Fuller Test

data:  log(PepperPrice[, "white"])
Dickey-Fuller = -1.744, Lag order = 6, p-value = 0.6838
alternative hypothesis: stationary
```

while, for the series of first differences, we obtain

```
R> adf.test(diff(log(PepperPrice[, "white"])))

        Augmented Dickey-Fuller Test

data:  diff(log(PepperPrice[, "white"]))
Dickey-Fuller = -5.336, Lag order = 6, p-value = 0.01
alternative hypothesis: stationary

Warning message:
In adf.test(diff(log(PepperPrice[, "white"]))) :
  p-value smaller than printed p-value
```

Note that a warning is issued because the p value is only interpolated from a few tabulated critical values, and hence no p values outside the interval $[0.01, 0.1]$ can be provided.

Alternatively, the Phillips-Perron test (Phillips and Perron 1988) with its nonparametric correction for autocorrelation (essentially employing a HAC estimate of the long-run variance in a Dickey-Fuller-type test (6.3) instead of parametric decorrelation) can be used. It is available in the function pp.test() from the package **tseries** (there also exists a function PP.test() in base R, but it has fewer options). Using the default options to the Phillips-Perron t test in the equation with a time trend, we obtain

```
R> pp.test(log(PepperPrice[, "white"]), type = "Z(t_alpha)")

        Phillips-Perron Unit Root Test

data:  log(PepperPrice[, "white"])
```

```
Dickey-Fuller Z(t_alpha) = -1.6439, Truncation lag
parameter = 5, p-value = 0.726
alternative hypothesis: stationary
```

Thus, all tests suggest that the null hypothesis of a unit root cannot be rejected here. Alternative implementations of the preceding methods with somewhat different interfaces are available in the package **urca** (Pfaff 2006). That package also offers a function ur.ers() implementing the test of Elliott, Rothenberg, and Stock (1996), which utilizes GLS detrending.

Stationarity tests

Kwiatkowski *et al.* (1992) proceed by testing for the presence of a random walk component r_t in the regression

$$y_t = d_t + r_t + \varepsilon_t,$$

where d_t denotes a deterministic component and ε_t is a stationary—more precisely, $I(0)$—error process. This test is also available in the function kpss.test() in the package **tseries**. The deterministic component is either a constant or a linear time trend, the former being the default. Setting the argument null = "Trend" yields the second version. Here, we obtain

```
R> kpss.test(log(PepperPrice[, "white"]))

        KPSS Test for Level Stationarity

data:  log(PepperPrice[, "white"])
KPSS Level = 0.9129, Truncation lag parameter = 3,
p-value = 0.01
```

Hence the KPSS test also points to nonstationarity of the pepper price series. (Again, a warning is issued, as the p value is interpolated from the four critical values provided by Kwiatkowski *et al.* 1992, it is suppressed here.)

Readers may want to check that the series pertaining to black pepper yields similar results when tested for unit roots or stationarity.

Cointegration

The very nature of the two pepper series already suggests that they possess common features. Having evidence for nonstationarity, it is of interest to test for a common nonstationary component by means of a cointegration test.

A simple method to test for cointegration is the two-step method proposed by Engle and Granger (1987). It regresses one series on the other and performs a unit root test on the residuals. This test, often named after Phillips and Ouliaris (1990), who provided the asymptotic theory, is available in the function po.test() from the package **tseries**. Specifically, po.test() performs a

Phillips-Perron test using an auxiliary regression without a constant and linear trend and the Newey-West estimator for the required long-run variance. A regression of the price for black pepper on that for white pepper yields

```
R> po.test(log(PepperPrice))
```

```
        Phillips-Ouliaris Cointegration Test
```

```
data:  log(PepperPrice)
Phillips-Ouliaris demeaned = -24.0987, Truncation lag
parameter = 2, p-value = 0.02404
```

suggesting that both series are cointegrated. (Recall that the first series pertains to black pepper. The function proceeds by regressing the first series on the remaining ones.) A test utilizing the reverse regression is as easy as po.test(log(PepperPrice[,2:1])). However, the problem with this approach is that it treats both series in an asymmetric fashion, while the concept of cointegration demands that the treatment be symmetric.

The standard tests proceeding in a symmetric manner stem from Johansen's full-information maximum likelihood approach (Johansen 1991). For a pth-order cointegrated vector autoregressive (VAR) model, the error correction form is (omitting deterministic components)

$$\Delta y_t = \Pi y_{t-1} + \sum_{j=1}^{p-1} \Gamma_j \Delta y_{t-j} + \varepsilon_t.$$

The relevant tests are available in the function ca.jo() from the package **urca**. The basic version considers the eigenvalues of the matrix Π in the preceding equation. We again refer to Hamilton (1994) for the methodological background.

Here, we employ the trace statistic—the maximum eigenvalue, or "lambda-max", test is available as well—in an equation amended by a constant term (specified by ecdet = "const"), yielding

```
R> library("urca")
R> pepper_jo <- ca.jo(log(PepperPrice), ecdet = "const",
+     type = "trace")
R> summary(pepper_jo)
```

```
######################
# Johansen-Procedure #
######################
```

```
Test type: trace statistic , without linear trend and
constant in cointegration
```

```
Eigenvalues (lambda):
```

[1] 4.93195e-02 1.35081e-02 1.38778e-17

Values of teststatistic and critical values of test:

```
          test 10pct  5pct  1pct
r <= 1 |  3.66   7.52   9.24 12.97
r = 0  | 17.26 17.85  19.96 24.60
```

Eigenvectors, normalised to first column:
(These are the cointegration relations)

```
            black.12 white.12   constant
black.12   1.000000  1.00000    1.00000
white.12  -0.889231 -5.09942    2.28091
constant  -0.556994 33.02742  -20.03244
```

Weights W:
(This is the loading matrix)

```
          black.12    white.12     constant
black.d  -0.0747230 0.00245321  3.86752e-17
white.d   0.0201561 0.00353701  4.03196e-18
```

The null hypothesis of no cointegration is rejected; hence the Johansen test confirms the results from the initial two-step approach.

6.4 Time Series Regression and Structural Change

More on fitting dynamic regression models

As already discussed in Chapter 3, there are various ways of fitting dynamic linear regression models by OLS in R. Here, we present two approaches in more detail: (1) setting up lagged and differenced regressors "by hand" and calling lm(); (2) using the convenience interface dynlm() from the package **dynlm** (Zeileis 2008). We illustrate both approaches using a model for the UKDriverDeaths series: the log-casualties are regressed on their lags 1 and 12, essentially corresponding to the multiplicative SARIMA$(1,0,0)(1,0,0)_{12}$ model

$$y_t = \beta_1 + \beta_2\, y_{t-1} + \beta_3\, y_{t-12} + \varepsilon_t, \quad t = 13, \ldots, 192.$$

For using lm() directly, we set up a multivariate time series containing the original log-casualties along with two further variables holding the lagged observations. The lagged variables are created with lag(). Note that its second argument, the number of lags, must be negative to shift back the observations. For "ts" series, this just amounts to changing the "tsp" attribute (leaving the

observations unchanged), whereas for "zoo" series k observations have to be
omitted for computation of the kth lag. For creating unions or intersections
of several "ts" series, ts.union() and ts.intersect() can be used, respectively. For "zoo" series, both operations are provided by the merge() method.
Here, we use ts.intersect() to combine the original and lagged series, assuring that leading and trailing NAs are omitted before the model fitting. The
final call to lm() works just as in the preceding sections because lm() does not
need to know that the underlying observations are from a single time series.

```
R> dd <- log(UKDriverDeaths)
R> dd_dat <- ts.intersect(dd, dd1 = lag(dd, k = -1),
+    dd12 = lag(dd, k = -12))
R> lm(dd ~ dd1 + dd12, data = dd_dat)

Call:
lm(formula = dd ~ dd1 + dd12, data = dd_dat)

Coefficients:
(Intercept)         dd1          dd12
      0.421         0.431         0.511
```

The disadvantage is that lm() cannot preserve time series properties of the
data unless further effort is made (specifically, setting dframe = TRUE in
ts.intersect() and na.action = NULL in lm(); see ?lm for details). Even
then, various nuisances remain, such as using different na.actions, print output formatting, or subset selection.

The function dynlm() addresses these issues. It provides an extended
model language in which differences and lags can be directly specified via
d() and L() (using the opposite sign of lag() for the second argument),
respectively. Thus

```
R> library("dynlm")
R> dynlm(dd ~ L(dd) + L(dd, 12))

Time series regression with "ts" data:
Start = 1970(1), End = 1984(12)

Call:
dynlm(formula = dd ~ L(dd) + L(dd, 12))

Coefficients:
(Intercept)        L(dd)      L(dd, 12)
      0.421        0.431          0.511
```

yields the same results as above, but the object returned is a "dynlm" object
inheriting from "lm" and provides additional information on the underlying
time stamps. The same model could be written somewhat more concisely as

dd ˜ L(dd, c(1, 12)). Currently, the disadvantage of dynlm() compared with lm() is that it cannot be reused as easily with other functions. However, as **dynlm** is still under development, this is likely to improve in future versions.

Structural change tests

As we have seen in previous sections, the structure in the series of log-casualties did not remain the same throughout the full sample period: there seemed to be a decrease in the mean number of casualties after the policy change in seatbelt legislation. Translated into a parametric time series model, this means that the parameters of the model are not stable throughout the sample period but change over time.

The package **strucchange** (Zeileis, Leisch, Hornik, and Kleiber 2002) implements a large collection of tests for structural change or parameter instability that can be broadly placed in two classes: (1) fluctuation tests and (2) tests based on F statistics. Fluctuation tests try to assess the structural stability by capturing fluctuation in cumulative or moving sums (CUSUMs or MOSUMs) of residuals (OLS or recursive), model scores (i.e., empirical estimating functions), or parameter estimates (from recursively growing or from rolling data windows). The idea is that, under the null hypothesis of parameter stability, the resulting "fluctuation processes" are governed by a functional central limit theorem and only exhibit limited fluctuation, whereas under the alternative of structural change, the fluctuation is generally increased. Thus, there is evidence for structural change if an appropriately chosen empirical fluctuation process crosses a boundary that the corresponding limiting process crosses only with probability α. In **strucchange**, empirical fluctuation processes can be fitted via efp(), returning an object of class "efp" that has a plot() method for performing the corresponding test graphically and an sctest() method (for structural change test) for a traditional significance test with test statistic and p value.

Here, we use an OLS-based CUSUM test (Ploberger and Krämer 1992) to assess the stability of the SARIMA-type model for the UK driver deaths data fitted at the beginning of this section. The OLS-based CUSUM process is simply the scaled cumulative sum process of the OLS residuals $\hat{\varepsilon}_t = y_t - x_t^\top \hat{\beta}$; that is,

$$efp(s) = \frac{1}{\hat{\sigma}\sqrt{n}} \sum_{t=1}^{\lfloor ns \rfloor} \hat{\varepsilon}_t, \quad 0 \le s \le 1.$$

It can be computed with the function efp() by supplying formula and data (as for lm()) and setting in addition type = "OLS-CUSUM":

```
R> dd_ocus <- efp(dd ˜ dd1 + dd12, data = dd_dat,
+     type = "OLS-CUSUM")
```

The associated structural change test, by default considering the maximum absolute deviation of the empirical fluctuation process from zero and given by

```
R> sctest(dd_ocus)
```

```
        OLS-based CUSUM test
```

```
data:   dd_ocus
S0 = 1.4866, p-value = 0.02407
```

is significant at the default 5% level, signaling that the model parameters are not stable throughout the entire sample period. The plot in the left panel of Figure 6.10 results from

```
R> plot(dd_ocus)
```

and yields some further insights. In addition to the excessive fluctuation (conveyed by the boundary crossing), it can be seen from the peak in the process that an abrupt change seems to have taken place in about 1973(10), matching the timing of the first oil crisis. A smaller second peak in the process, associated with the change of seatbelt legislation in 1983(1), is also visible.

Tests based on F statistics, the second class of tests in **strucchange**, are designed to have good power for single-shift alternatives (of unknown timing). The basic idea is to compute an F statistic (or Chow statistic) for each conceivable breakpoint in a certain interval and reject the null hypothesis of structural stability if any of these statistics (or some other functional such as the mean) exceeds a certain critical value (Andrews 1993; Andrews and Ploberger 1994). Processes of F statistics can be fitted with Fstats(), employing an interface similar to efp(). The resulting "Fstats" objects can again be assessed by the corresponding sctest() method or graphically by the plot() method. The code chunk

```
R> dd_fs <- Fstats(dd ~ dd1 + dd12, data = dd_dat, from = 0.1)
R> plot(dd_fs)
R> sctest(dd_fs)
```

```
        supF test
```

```
data:   dd_fs
sup.F = 19.3331, p-value = 0.006721
```

uses the sup$F$ test of Andrews (1993) for the SARIMA-type model with a trimming of 10%; i.e., an F statistic is computed for each potential breakpoint between 1971(6) and 1983(6), omitting the leading and trailing 10% of observations. The resulting process of F statistics is shown in the right panel of Figure 6.10, revealing two clear peaks in 1973(10) and 1983(1). Both the boundary crossing and the tiny p value show that there is significant departure from the null hypothesis of structural stability. The two peaks in the F process also demonstrate that although designed for single-shift alternatives, the sup$F$ test has power against multiple-shift alternatives. In this case, it brings out the two breaks even more clearly than the OLS-based CUSUM test.

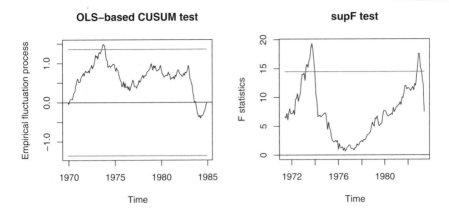

Fig. 6.10. OLS-based CUSUM process (left) and process of F statistics (right) for the UK driver deaths model.

Many further tests for structural change are available in **strucchange**. A unified and more flexible approach is implemented in `gefp()` (Zeileis 2005, 2006a). In practice, this multitude of tests is often a curse rather than a blessing. Unfortunately, no test is superior to any other test for all conceivable patterns of structural change. Hence, the choice of a suitable test can be facilitated if there is some prior knowledge about which types of changes are likely to occur and which parameters are affected by it (see Zeileis 2005, for some discussion of this).

To further illustrate the wide variety of structural change tests, we consider a second example. Lütkepohl, Teräsvirta, and Wolters (1999) establish an error correction model (ECM) for German M1 money demand, reanalyzed by Zeileis, Leisch, Kleiber, and Hornik (2005) in a structural change setting. The data frame `GermanM1` contains data from 1961(1) to 1995(4) on per capita M1, price index, per capita GNP (all in logs) and an interest rate. It can be loaded along with the model used by Lütkepohl *et al.* (1999) via

```
R> data("GermanM1")
R> LTW <- dm ~ dy2 + dR + dR1 + dp + m1 + y1 + R1 + season
```

involving the differenced and lagged series as well as a factor `season` that codes the quarter of the year. To test whether a stable model can be fitted for this ECM, particularly in view of the German monetary unification on 1990-06-01, a recursive estimates (RE) test (the "fluctuation test" of Ploberger, Krämer, and Kontrus 1989), is employed using

```
R> m1_re <- efp(LTW, data = GermanM1, type = "RE")
R> plot(m1_re)
```

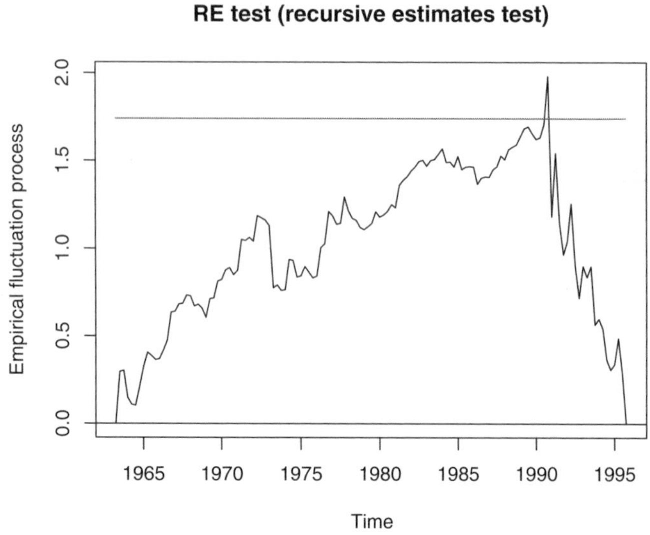

RE test (recursive estimates test)

Fig. 6.11. Recursive estimates fluctuation process for German M1 model.

The boundary crossing of the RE process shown in Figure 6.11 signals again that there is a deviation from structural stability (at the default 5% level), and the clear peak conveys that this is due to an abrupt change in 1990, matching the timing of the German monetary unification.

Dating structural changes

Given that there is evidence for structural change in a certain model, a natural strategy is to find a model that incorporates the changes. When the changes are abrupt, this amounts to segmenting the original data and fitting the model on each subset. In the framework of the linear regression model, the setup is

$$y_t = x_t^\top \beta^{(j)} + \varepsilon_t, \qquad t = n_{j-1} + 1, \ldots, n_j, \quad j = 1, \ldots, m + 1, \qquad (6.4)$$

where $j = 1, \ldots, m$ is the segment index and $\beta^{(j)}$ is the segment-specific set of regression coefficients. The indices $\{n_1, \ldots, n_m\}$ denote the set of unknown breakpoints, and by convention $n_0 = 0$ and $n_{m+1} = n$.

Estimating the breakpoints is also called *dating* structural changes. For the two models considered above, visual inspection already provides information on the locations of the breakpoints. However, a more formal procedure for determining the number and location of breakpoints is desirable. Bai and Perron (1998, 2003) established a general methodology for estimating breakpoints and their associated confidence intervals in OLS regression, and their method is implemented in the function breakpoints() by Zeileis, Kleiber, Krämer, and Hornik (2003). The dating procedure of Bai and Perron (2003) employs a dynamic programming algorithm based on the Bellman principle for finding those m breakpoints that minimize the residual sum of squares (RSS) of a model with $m + 1$ segments, given some minimal segment size of $h \cdot n$ observations. Here, h is a bandwidth parameter to be chosen by the user. Similar to the choice of trimming for the F statistics-based tests, the minimal proportion of observations in each segment is typically chosen to be 10% or 15%. Given h and m, the breakpoints minimizing the RSS can be determined; however, typically the number of breakpoints m is not known in advance. One possibility is to compute the optimal breakpoints for $m = 0, 1, \ldots$ breaks and choose the model that minimizes some information criterion such as the BIC. This model selection strategy is also directly available within breakpoints().

Returning to the UKDriverDeaths series, we estimate the breakpoints for a SARIMA-type model with a minimal segment size of 10% using

```r
R> dd_bp <- breakpoints(dd ~ dd1 + dd12, data = dd_dat, h = 0.1)
```

The RSS and BIC as displayed by plot(dd_bp) are shown in the left panel of Figure 6.12. Although the RSS drops clearly up to $m = 3$ breaks, the BIC is minimal for $m = 0$ breaks. This is not very satisfactory, as the structural change tests clearly showed that the model parameters are not stable. As the BIC was found to be somewhat unreliable for autoregressive models by Bai and Perron (2003), we rely on the interpretation from the visualization of the structural change tests and use the model with $m = 2$ breaks. Its coefficients can be extracted via

```r
R> coef(dd_bp, breaks = 2)
```

	(Intercept)	dd1	dd12
1970(1) - 1973(10)	1.45776	0.117323	0.694480
1973(11) - 1983(1)	1.53421	0.218214	0.572330
1983(2) - 1984(12)	1.68690	0.548609	0.214166

reflecting that particularly the period after the change in seatbelt legislation in 1983(1) is different from the observations before. The other breakpoint is in 1973(10), again matching the timing of the oil crisis and confirming the interpretation from the structural change tests. The observed and fitted series, along with confidence intervals for the breakpoints, are shown in the right panel of Figure 6.12 as generated by

```r
R> plot(dd)
R> lines(fitted(dd_bp, breaks = 2), col = 4)
R> lines(confint(dd_bp, breaks = 2))
```

Readers are asked to estimate breakpoints for the GermanM1 example in an exercise.

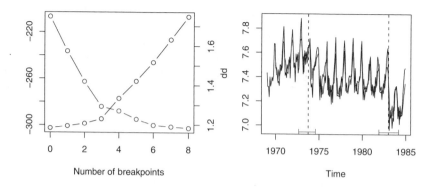

Fig. 6.12. Left panel: BIC and residual sum of squares for segmented UK driver death models. Right panel: observed and fitted segmented series.

6.5 Extensions

For reasons of space, our discussion of time series methods has been rather brief. Table 6.3 provides a list of several further packages containing time series functions.

The remainder of this section briefly considers structural time series models and the most widely used volatility model, the GARCH(1,1).

Structural time series models

Structural time series models are state-space models utilizing a decomposition of the time series into a number of components that are specified by a set of disturbance variances. Thus, these models may be considered error component models for time series data. Harvey (1989) and Durbin and Koopman (2001) are standard references.

StructTS() from the package **stats** fits a subclass of (linear and Gaussian) state-space models, including the so-called *basic structural model* defined via the measurement equation

$$y_t = \mu_t + \gamma_t + \varepsilon_t, \quad \varepsilon_t \sim \mathcal{N}(0, \sigma_\varepsilon^2) \text{ i.i.d.,}$$

where γ_t is a seasonal component (with frequency s) defined as $\gamma_{t+1} = -\sum_{j=1}^{s-1} \gamma_{t+1-j} + \omega_t$, $\omega_t \sim \mathcal{N}(0, \sigma_\omega^2)$ i.i.d., and the local level and trend components are given by

$$\mu_{t+1} = \mu_t + \eta_t + \xi_t, \quad \xi_t \sim \mathcal{N}(0, \sigma_\xi^2) \text{ i.i.d.,}$$
$$\eta_{t+1} = \eta_t + \zeta_t, \quad \zeta_t \sim \mathcal{N}(0, \sigma_\zeta^2) \text{ i.i.d.}$$

Table 6.3. Further packages for time series analysis.

Package	Description
dse	Multivariate time series modeling with state-space and vector ARMA (VARMA) models (Gilbert 2007).
FinTS	R companion to Tsay (2005) with data sets, functions, and script files to work some of the examples (Graves 2008).
forecast	Univariate time series forecasting, including exponential smoothing, state space, and ARIMA models. Part of the **forecasting** bundle (Hyndman and Khandakar 2008).
fracdiff	ML estimation of fractionally integrated ARMA (ARFIMA) models and semiparametric estimation of the fractional differencing parameter (Fraley, Leisch, and Maechler 2006).
longmemo	Convenience functions for long-memory models; also contains several data sets (Beran, Whitcher, and Maechler 2007).
mFilter	Miscellaneous time series filters, including Baxter-King, Butterworth, and Hodrick-Prescott (Balcilar 2007).
Rmetrics	Suite of some 20 packages for financial engineering and computational finance (Wuertz 2008), including GARCH modeling in the package **fGarch**.
tsDyn	Nonlinear time series models: STAR, ESTAR, LSTAR (Di Narzo and Aznarte 2008).
vars	(Structural) vector autoregressive (VAR) models (Pfaff 2008).

All error terms are assumed to be mutually independent. Special cases are the local linear trend model, where γ_t is absent, and the local linear model, where in addition $\sigma_\zeta^2 = 0$.

In total, there are four parameters, σ_ξ^2, σ_η^2, σ_ω^2, and σ_ε^2, some (but not all) of which may be absent (and often are in practical applications).

It should be noted that, for example, the reduced form of the local trend model is ARIMA(0,2,2), but with restrictions on the parameter set. Proponents of structural time series models argue that the implied restrictions often are meaningful in practical terms and thus lead to models that are easier to interpret than results from unrestricted ARIMA fits.

Here, we fit the basic structural model to the UKDriverDeaths data using

```
R> dd_struct <- StructTS(log(UKDriverDeaths))
```

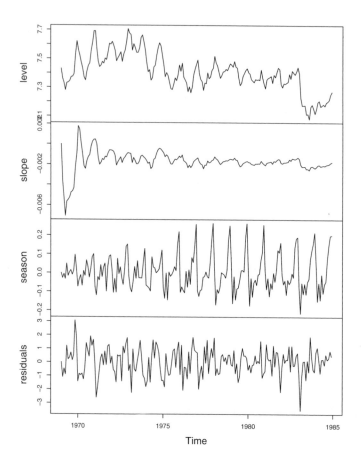

Fig. 6.13. The basic structural model for the UK driver deaths.

The resulting components

```
R> plot(cbind(fitted(dd_struct), residuals(dd_struct)))
```

are shown in Figure 6.13. This approach also clearly brings out the drop in the number of accidents in connection with the change in legislation.

More information on structural time series models in R is provided by Ripley (2002) and Venables and Ripley (2002).

GARCH models

Many financial time series exhibit volatility clustering. Figure 6.14 provides a typical example, a series of 1974 DEM/GBP exchange rate returns for the

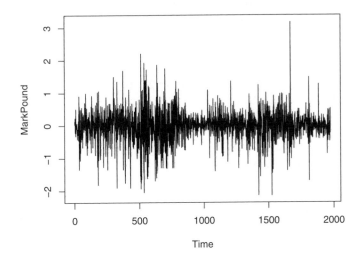

Fig. 6.14. DEM/GBP exchange rate returns.

period 1984-01-03 through 1991-12-31, taken from Bollerslev and Ghysels (1996). This data set has recently become a benchmark in the GARCH literature, and it is also used in Greene (2003, Chapter 11). A Ljung-Box test of the MarkPound series suggests that it is white noise (note that this is not without problems since this test assumes i.i.d. data under the null hypothesis), while a test of the squares is highly significant. Thus, a GARCH model might be appropriate.

The package **tseries** provides a function garch() for the fitting of GARCH(p, q) models with Gaussian innovations, defaulting to the popular GARCH$(1, 1)$

$$y_t = \sigma_t \nu_t, \quad \nu_t \sim \mathcal{N}(0, 1) \text{ i.i.d.,}$$
$$\sigma_t^2 = \omega + \alpha y_{t-1}^2 + \beta \sigma_{t-1}^2, \quad \omega > 0, \alpha > 0, \beta \geq 0.$$

For the exchange rate data, we obtain

```
R> mp <- garch(MarkPound, grad = "numerical", trace = FALSE)
R> summary(mp)

Call:
garch(x = MarkPound, grad = "numerical", trace = FALSE)

Model:
GARCH(1,1)
```

```
Residuals:
     Min        1Q     Median        3Q       Max
-6.79739 -0.53703 -0.00264   0.55233   5.24867

Coefficient(s):
    Estimate  Std. Error  t value  Pr(>|t|)
a0    0.0109      0.0013     8.38    <2e-16
a1    0.1546      0.0139    11.14    <2e-16
b1    0.8044      0.0160    50.13    <2e-16

Diagnostic Tests:
         Jarque Bera Test

data:  Residuals
X-squared = 1060.01, df = 2, p-value < 2.2e-16


         Box-Ljung test

data:  Squared.Residuals
X-squared = 2.4776, df = 1, p-value = 0.1155
```

which gives ML estimates along with outer-product-of-gradients (OPG) standard errors and also reports diagnostic tests of the residuals for normality (rejected) and independence (not rejected). Numerical (rather than the default analytical) gradients are employed in the `garch()` call because the resulting maximized log-likelihood is slightly larger. For brevity, `trace = FALSE` suppresses convergence diagnostics, but we encourage readers to reenable this option in applications.

More elaborate tools for volatility modeling are available in the **Rmetrics** collection of packages (Wuertz 2008). We mention the function `garchFit()` from the package **fGarch**, which includes, among other features, a wider selection of innovation distributions in GARCH specifications.

6.6 Exercises

1. Decompose the `UKNonDurables` data. Filter these data using the Holt-Winters technique.
2. Repeat Exercise 1 for the `DutchSales` data. Compare a decomposition via `decompose()` with a decomposition via `stl()`.
3. Using the `AirPassengers` series,
   - filter the data using Holt-Winters smoothing,
   - fit the airline model, and

- fit the basic structural model.

Compare the results.

4. Reread Nelson and Plosser (1982) and test a selection of the extended Nelson-Plosser data, available as `NelPlo` in the package **tseries**, for unit roots. Also fit ARIMA models to these series. (You may need the `xreg` argument to `arima()` for series that appear to be trend stationary.)

5. Compute and plot the coefficients of the implied $MA(\infty)$ representations (i.e., the impulse response functions) for (the stationary part of) the models fitted in the preceding exercise.

   Hint: Use `ARMAtoMA()`.

6. Stock and Watson (2007) consider an autoregressive distributed lag model for the change in the US inflation rate using the `USMacroSW` data. Specifically, the model is

$$\Delta \mathtt{inf}_t = \beta_0 + \sum_{i=1}^{4} \beta_i \Delta \mathtt{inf}_{t-i} + \sum_{j=1}^{4} \gamma_j \mathtt{unemp}_{t-j} + \varepsilon_t.$$

   Plot the sequence of F statistics for a single structural break for this $ADL(4, 4)$ model using `Fstats()` and test for structural changes with the $\sup F$ test.

7. Apply the Bai and Perron (2003) dating algorithm to the German M1 data. Do the results correspond to the German monetary reunion?

8. Fit the basic structural model to the `UKNonDurables` data. Compare this with the ARIMA model fitted in Section 6.2.

7

Programming Your Own Analysis

Data analysis, both in academic and in corporate environments, typically involves, in some form or other, the following three components: (a) using or writing software that can perform the desired analysis, (b) a sequence of commands or instructions that apply the software to the data set under investigation, and (c) documentation of the commands and their output.

R comes with a rich suite of tools that help implement all these steps while making the analysis reproducible and applicable to new data. So far, we have mostly been concerned with providing short examples of existing functionality. In this chapter, we try to enrich this picture by illustrating how further aspects of the tasks above can be performed:

(a) In the simplest case, a function that performs exactly the analysis desired is already available. This is the case for many standard models, as discussed in the preceding chapters. In the worst case, no infrastructure is available yet, and a new function has to be written from scratch. In most cases, however, something in between is required: a (set of) new function(s) that reuse(s) existing functionality. This can range from simple convenience interfaces to extensions built on top of existing functions. In any case, the resulting software is most easily applicable if the functions reflect the conceptual steps in the analysis.

(b) As R is mostly used via its command-line interface, assembling scripts with R commands is straightforward (and the `history()` from an R session is often a good starting point). To make the results reproducible, it is important to keep track of the entire analysis, from data preprocessing over model fitting to evaluation and generation of output and graphics.

(c) For documenting results obtained from an analysis in R, many environments are available, ranging from word processors to markup languages such as HTML or LATEX. For all of these it is possible to produce R output—numerical and/or graphical—and to include this "by hand" in the documentation (e.g., by "copy and paste"). However, this can be tedious and, more importantly, make replication or application to a different data

C. Kleiber, A. Zeileis, *Applied Econometrics with R*,
DOI: 10.1007/978-0-387-77318-6_7, © Springer Science+Business Media, LLC 2008

set much more cumbersome. Therefore, R ships with support for tightly bundling R scripts (as discussed in (b)) and the documentation of their output so that R first runs the analysis and then includes the results in the documentation. Base R provides the function `Sweave()` (in package **utils**), which by default supports "weaving" of R code with LaTeX documentation but also allows other documentation formats to be plugged in.

In the following, we provide examples for a few typical tasks in econometric analysis—simulation of power curves, bootstrapping a regression, and maximizing a likelihood. In doing so, we go beyond using off-the-shelf software and in each case require some of the steps discussed above.

7.1 Simulations

A simulation study is one of the most typical programming tasks when evaluating some algorithm; e.g., a test procedure or an estimator. It usually involves (at least) three steps: (1) simulating data from some data-generating process (DGP); (2) evaluating the quantities of interest (e.g., rejection probabilities, parameter estimates, model predictions); and (3) iterating the first two steps over a number of different scenarios. Here, we exemplify how to accomplish such a task in R by comparing the power of two well-known tests for autocorrelation—the Durbin-Watson and the Breusch-Godfrey test—in two different specifications of a linear regression. In the following, we first set up three functions that capture the steps above before we actually run the simulation and summarize the results both numerically and graphically.

Data-generating process

We consider the Durbin-Watson and Breusch-Godfrey tests for two different linear regression models: a trend model with regressor $x_i = i$ and a model with a lagged dependent variable $x_i = y_{i-1}$. Recall that the Durbin-Watson test is not valid in the presence of lagged dependent variables.

More specifically, the model equations are

$$\text{trend:} \quad y_i = \beta_1 + \beta_2 \cdot i + \varepsilon_i,$$
$$\text{dynamic:} \quad y_i = \beta_1 + \beta_2 \cdot y_{i-1} + \varepsilon_i,$$

where the regression coefficients are in both cases $\beta = (0.25, -0.75)^\top$, and $\{\varepsilon_i\}$, $i = 1, \ldots, n$, is a stationary AR(1) series, derived from standard normal innovations and with lag 1 autocorrelation ϱ. All starting values, both for y and ε, are chosen as 0.

We want to analyze the power properties of the two tests (for size $\alpha = 0.05$) on the two DGPs for autocorrelations $\varrho = 0, 0.2, 0.4, 0.6, 0.8, 0.9, 0.95, 0.99$ and sample sizes $n = 15, 30, 50$.

To carry out such a simulation in R, we first define a function `dgp()` that implements the DGP above:

```
R> dgp <- function(nobs = 15, model = c("trend", "dynamic"),
+    corr = 0, coef = c(0.25, -0.75), sd = 1)
+  {
+    model <- match.arg(model)
+    coef <- rep(coef, length.out = 2)
+
+    err <- as.vector(filter(rnorm(nobs, sd = sd), corr,
+      method = "recursive"))
+    if(model == "trend") {
+      x <- 1:nobs
+      y <- coef[1] + coef[2] * x + err
+    } else {
+      y <- rep(NA, nobs)
+      y[1] <- coef[1] + err[1]
+      for(i in 2:nobs)
+        y[i] <- coef[1] + coef[2] * y[i-1] + err[i]
+      x <- c(0, y[1:(nobs-1)])
+    }
+    return(data.frame(y = y, x = x))
+  }
```

The arguments to dgp() are nobs (corresponding to n with default 15), model (specifying the equation used, by default "trend"), corr (the autocorrelation ϱ, by default 0), coef (corresponding to β), and sd (the standard deviation of the innovation series). The latter two are held constant in the following. After assuring that model and coef are of the form required, dgp() sets up the regressor and dependent variable and returns both in a "data.frame" comprising the variables y and x.

Evaluation for a single scenario

Based on this implementation of the DGP, we can now easily simulate the power of both tests for a given combination of parameters. In simpower(), we just iterate through a for() loop (by default with nrep = 100 iterations) in which we simulate a data set, apply both dwtest() (the Durbin-Watson test from **lmtest**) and bgtest() (the Breusch-Godfrey test from **lmtest**) to it, and store the associated p values. After completing the for() loop, we return the proportion of significant p values (by default at size = 0.05).

```
R> simpower <- function(nrep = 100, size = 0.05, ...)
+  {
+    pval <- matrix(rep(NA, 2 * nrep), ncol = 2)
+    colnames(pval) <- c("dwtest", "bgtest")
+    for(i in 1:nrep) {
+      dat <- dgp(...)
+      pval[i,1] <- dwtest(y ~ x, data = dat,
```

```
+        alternative = "two.sided")$p.value
+      pval[i,2] <- bgtest(y ~ x, data = dat)$p.value
+    }
+    return(colMeans(pval < size))
+ }
```

The remaining argument ... is a simple mechanism for passing on further arguments to other functions. Here, any argument beyond `nrep` and `size` is passed to `dgp()`. For example, we could easily modify the correlation employed via `simpower(corr = 0.9)`, which would then internally call `dgp(corr = 0.9)`, all other defaults remaining unchanged.

Iterated evaluation over all scenarios

Given these two building blocks, the DGP and the power simulator, we can now set up the main simulation routine in which we compute the power for different combinations of autocorrelation, sample size, and regression model:

```
R> simulation <- function(corr = c(0, 0.2, 0.4, 0.6, 0.8,
+    0.9, 0.95, 0.99), nobs = c(15, 30, 50),
+    model = c("trend", "dynamic"), ...)
+ {
+    prs <- expand.grid(corr = corr, nobs = nobs, model = model)
+    nprs <- nrow(prs)
+
+    pow <- matrix(rep(NA, 2 * nprs), ncol = 2)
+    for(i in 1:nprs) pow[i,] <- simpower(corr = prs[i,1],
+      nobs = prs[i,2], model = as.character(prs[i,3]), ...)
+
+    rval <- rbind(prs, prs)
+    rval$test <- factor(rep(1:2, c(nprs, nprs)),
+      labels = c("dwtest", "bgtest"))
+    rval$power <- c(pow[,1], pow[,2])
+    rval$nobs <- factor(rval$nobs)
+    return(rval)
+ }
```

This function simply sets up all parameter combinations in a "data.frame" using `expand.grid()` and subsequently simulates both power values for each of the parameter combinations in a `for()` loop. Finally, the results are slightly rearranged and returned in a "data.frame".

Of course, it would have been possible to code all the preceding steps in a single function; however, such functions tend to be rather monolithic and not very intelligible. Encapsulating logical steps in computational building blocks helps to make simulation code readable and facilitates reuse outside a large simulation. This is particularly helpful during the setup phase, where sanity checking of the building blocks is possible independently.

Simulation and summary

Now, all that is required to run the simulation is to set a random seed (so that the simulation results can always be exactly reproduced) and call simulation():

```
R> set.seed(123)
R> psim <- simulation()
```

Using the default settings, this takes less than a minute on a standard PC; however, the precision from only 100 replications is certainly not sufficient for professional applications.

To inspect the simulation results, the most standard and simple format is tables with numerical output. Using xtabs(), we can turn the "data.frame" into a "table" that classifies the power outcome by the four design variables. For printing the resulting four-way table, we create a "flat" two-way table. This can be achieved using ftable() (for flat table). Placing the values for ϱ in the columns and nesting all other variables in the rows, we obtain

```
R> tab <- xtabs(power ~ corr + test + model + nobs, data = psim)
R> ftable(tab, row.vars = c("model", "nobs", "test"),
+     col.vars = "corr")
```

model	nobs	test	corr	0	0.2	0.4	0.6	0.8	0.9	0.95	0.99
trend	15	dwtest		0.05	0.10	0.21	0.36	0.55	0.65	0.66	0.62
		bgtest		0.07	0.05	0.05	0.10	0.30	0.40	0.41	0.31
	30	dwtest		0.09	0.20	0.57	0.80	0.96	1.00	0.96	0.98
		bgtest		0.09	0.09	0.37	0.69	0.93	0.99	0.94	0.93
	50	dwtest		0.03	0.31	0.76	0.99	1.00	1.00	1.00	1.00
		bgtest		0.05	0.23	0.63	0.95	1.00	1.00	1.00	1.00
dynamic	15	dwtest		0.02	0.01	0.00	0.00	0.01	0.03	0.01	0.00
		bgtest		0.07	0.04	0.01	0.09	0.14	0.21	0.17	0.26
	30	dwtest		0.00	0.01	0.01	0.06	0.00	0.03	0.03	0.19
		bgtest		0.05	0.05	0.18	0.39	0.52	0.63	0.64	0.74
	50	dwtest		0.02	0.02	0.01	0.03	0.03	0.15	0.39	0.56
		bgtest		0.05	0.10	0.36	0.72	0.91	0.90	0.93	0.91

By supplying the test as the last row variable, the table is aimed at comparing the power curves (i.e., the rejection probabilities for increasing ϱ) between the two tests under investigation. It can be seen that the Durbin-Watson test performs somewhat better in the trend model, although the advantage over the Breusch-Godfrey test diminishes with increasing ϱ and n. As expected, for the dynamic model, the Durbin-Watson test has almost no power except for very high correlations whereas the Breusch-Godfrey test performs acceptably.

This difference becomes even more apparent when the comparison is carried out graphically. Instead of the standard R graphics, we prefer to use so-called trellis graphics for this task. R provides the package **lattice** (Sarkar

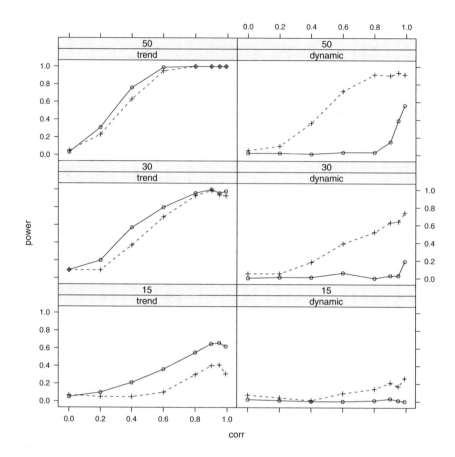

Fig. 7.1. Simulated power curves for `dwtest()` (solid) and `bgtest()` (dashed).

2002), specifically aimed at such layouts. It is written in the **grid** graphics system (Murrell 2005), a system even more flexible than R's default graphics facilities. **grid** comes with a multitude of functions and parameters for controlling possibly complex trellis graphics in publication quality. Here, we do not discuss in detail how to use **lattice** and **grid** (the interested reader is referred to the associated package documentation) and only demonstrate how to generate Figure 7.1:

```
R> library("lattice")
R> xyplot(power ~ corr | model + nobs, groups = ~ test,
+    data = psim, type = "b")
```

Using `xyplot()`, a trellis scatterplot is generated for `power ~ corr` conditional on the combinations of `model` and `nobs`. Within each panel, the observations are grouped by `test`. All data are taken from the simulation results `psim`, and the plotting type is `"b"`, indicating both (i.e., lines and points). For further control options, see `?xyplot`.

In Figure 7.1, rows correspond to varying sample sizes and columns to the underlying model, and each panel shows the power curves as functions of ϱ for both tests. The interpretation is much easier compared to the numerical table: power clearly increases with n and ϱ, and autocorrelation is easier to detect in the trend than in the dynamic model. While the Durbin-Watson test performs slightly better in the trend model for small sample sizes, its power breaks down almost completely in the dynamic model.

7.2 Bootstrapping a Linear Regression

Conventional regression output relies on asymptotic approximations to the distributions of test statistics, which are often not very reliable in small samples or models with substantial nonlinearities. A possible remedy is to employ bootstrap methodology.

In R, a basic recommended package is **boot** (Canty and Ripley 2008), which provides functions and data sets from Davison and Hinkley (1997). Specifically, the function `boot()` implements the classical nonparametric bootstrap (sampling with replacement), among other resampling techniques.

Since there is no such thing as "the" bootstrap, the first question is to determine a resampling strategy appropriate for the problem at hand. In econometrics and the social sciences, experimental data are rare, and hence it is appropriate to consider regressors as well as responses as random variables. This suggests employing the pairs bootstrap (i.e., to resample observations), a method that should give reliable standard errors even in the presence of (conditional) heteroskedasticity.

As an illustration, we revisit an example from Chapter 3, the `Journals` data taken from Stock and Watson (2007). The goal is to compute bootstrap standard errors and confidence intervals by case-based resampling. For ease of reference, we reproduce the basic regression

```
R> data("Journals")
R> journals <- Journals[, c("subs", "price")]
R> journals$citeprice <- Journals$price/Journals$citations
R> jour_lm <- lm(log(subs) ~ log(citeprice), data = journals)
```

The function `boot()` takes several arguments, of which `data`, `statistic`, and `R` are required. Here, `data` is simply the data set and `R` signifies the number of bootstrap replicates. The second argument, `statistic`, is a function that returns the statistic to be bootstrapped, where the function itself must take the data set and an index vector providing the indices of the observations included in the current bootstrap sample.

This is best understood by considering an example. In our case, the required statistic is given by the convenience function

```
R> refit <- function(data, i)
+    coef(lm(log(subs) ~ log(citeprice), data = data[i,]))
```

Now we are ready to call boot():

```
R> library("boot")
R> set.seed(123)
R> jour_boot <- boot(journals, refit, R = 999)
```

This yields

```
R> jour_boot
```

```
ORDINARY NONPARAMETRIC BOOTSTRAP


Call:
boot(data = journals, statistic = refit, R = 999)


Bootstrap Statistics :
     original       bias    std. error
t1*    4.7662 -0.0010560      0.05545
t2*   -0.5331 -0.0001606      0.03304
```

A comparison with the standard regression output

```
R> coeftest(jour_lm)
```

```
t test of coefficients:

                Estimate Std. Error t value Pr(>|t|)
(Intercept)       4.7662     0.0559    85.2   <2e-16
log(citeprice)   -0.5331     0.0356   -15.0   <2e-16
```

reveals only minor differences, suggesting that the conventional version is fairly reliable in this application.

We can also compare bootstrap and conventional confidence intervals. As the bootstrap standard errors were quite similar to the conventional ones, confidence intervals will be expected to be quite similar as well in this example. To save space, we confine ourselves to those of the slope coefficient. The bootstrap provides the interval

```
R> boot.ci(jour_boot, index = 2, type = "basic")
```

```
BOOTSTRAP CONFIDENCE INTERVAL CALCULATIONS
Based on 999 bootstrap replicates

CALL :
boot.ci(boot.out = jour_boot, type = "basic", index = 2)

Intervals :
Level      Basic
95%    (-0.5952, -0.4665 )
Calculations and Intervals on Original Scale
```

while its classical counterpart is

```
R> confint(jour_lm,  parm = 2)

                2.5 %  97.5 %
log(citeprice) -0.6033 -0.4628
```

This underlines that both approaches yield essentially identical results here. Note that `boot.ci()` provides several further methods for computing bootstrap intervals.

The bootstrap is particularly helpful in situations going beyond least-squares regression; thus readers are asked to explore bootstrap standard errors in connection with robust regression techniques in an exercise.

Finally, it should be noted that **boot** contains many further functions for resampling, among them `tsboot()` for block resampling from time series (for blocks of both fixed and random lengths). Similar functionality is provided by the function `tsbootstrap()` in the package **tseries**. In addition, a rather different approach, the maximum entropy bootstrap (Vinod 2006), is available in the package **meboot**.

7.3 Maximizing a Likelihood

Transformations of dependent variables are a popular means to improve the performance of models and are also helpful in the interpretation of results. Zellner and Revankar (1969), in a search for a generalized production function that allows returns to scale to vary with the level of output, introduced (among more general specifications) the generalized Cobb-Douglas form

$$Y_i e^{\theta Y_i} = e^{\beta_1} K_i^{\beta_2} L_i^{\beta_3},$$

where Y is output, K is capital, and L is labor. From a statistical point of view, this can be seen as a transformation applied to the dependent variable encompassing the level (for $\theta = 0$, which in this application yields the classical Cobb-Douglas function). Introducing a multiplicative error leads to the logarithmic form

$$\log Y_i + \theta Y_i = \beta_1 + \beta_2 \log K_i + \beta_3 \log L_i + \varepsilon_i. \tag{7.1}$$

However, this model is nonlinear in the parameters, and only for known θ can it be estimated by OLS. Following Zellner and Ryu (1998) and Greene (2003, Chapter 17), using the `Equipment` data on transportation equipment manufacturing, we attempt to simultaneously estimate the regression coefficients and the transformation parameter using maximum likelihood assuming $\varepsilon_i \sim \mathcal{N}(0, \sigma^2)$ i.i.d.

The likelihood of the model is

$$\mathcal{L} = \prod_{i=1}^{n} \left\{ \phi(\varepsilon_i/\sigma) \cdot \frac{1 + \theta Y_i}{Y_i} \right\},$$

where $\varepsilon_i = \log Y_i + \theta Y_i - \beta_1 - \beta_2 \log K_i - \beta_3 \log L_i$ and $\phi(\cdot)$ is the probability density function of the standard normal distribution. Note that $\partial \varepsilon_i / \partial Y_i = (1 + \theta Y_i)/Y_i$.

This gives the log-likelihood

$$\ell = \sum_{i=1}^{n} \left\{ \log(1 + \theta Y_i) - \log Y_i \right\} - \sum_{i=1}^{n} \log \phi(\varepsilon_i/\sigma).$$

The task is to write a function maximizing this log-likelihood with respect to the parameter vector $(\beta_1, \beta_2, \beta_3, \theta, \sigma^2)$. This decomposes naturally into the following three steps: (1) code the objective function, (2) obtain starting values for an iterative optimization, and (3) optimize the objective function using the starting values.

Step 1: We begin by coding the log-likelihood. However, since the function `optim()` used below by default performs minimization, we have to slightly modify the natural approach in that we need to *minimize* the *negative* of the log-likelihood.

```r
R> data("Equipment", package = "AER")
R> nlogL <- function(par) {
+     beta <- par[1:3]
+     theta <- par[4]
+     sigma2 <- par[5]
+
+     Y <- with(Equipment, valueadded/firms)
+     K <- with(Equipment, capital/firms)
+     L <- with(Equipment, labor/firms)
+
+     rhs <- beta[1] + beta[2] * log(K) + beta[3] * log(L)
+     lhs <- log(Y) + theta * Y
+
+     rval <- sum(log(1 + theta * Y) - log(Y) +
+         dnorm(lhs, mean = rhs, sd = sqrt(sigma2), log = TRUE))
```

```
+      return(-rval)
+   }
```

The function `nlogL()` is a function of a vector parameter `par` comprising five elements; for convenience, these are labeled as in Equation (7.1). Variables are transformed as needed, after which both sides of Equation (7.1) are set up. These ingredients are then used in the objective function `rval`, the negative of which is finally returned. Note that R comes with functions for the logarithms of the standard distributions, including the normal density `dnorm(..., log = TRUE)`.

Step 2: `optim()` proceeds iteratively, and thus (good) starting values are needed. These can be obtained from fitting the classical Cobb-Douglas form by OLS:

```
R> fm0 <- lm(log(valueadded/firms) ~ log(capital/firms) +
+      log(labor/firms), data = Equipment)
```

The resulting vector of coefficients, `coef(fm0)`, is now amended by 0, our starting value for θ, and the mean of the squared residuals from the Cobb-Douglas fit, the starting value for the disturbance variance:

```
R> par0 <- as.vector(c(coef(fm0), 0, mean(residuals(fm0)^2)))
```

Step 3: We are now ready to search for the optimum. The new vector `par0` containing all the starting values is used in the call to `optim()`:

```
R> opt <- optim(par0, nlogL, hessian = TRUE)
```

By default, `optim()` uses the Nelder-Mead method, but there are further algorithms available. We set `hessian = TRUE` in order to obtain standard errors. Parameter estimates, standard errors, and the value of the objective function at the estimates can now be extracted via

```
R> opt$par
```

```
[1] 2.91469 0.34998 1.09232 0.10666 0.04275
```

```
R> sqrt(diag(solve(opt$hessian)))[1:4]
```

```
[1] 0.36055 0.09671 0.14079 0.05850
```

```
R> -opt$value
```

```
[1] -8.939
```

In spite of the small sample, these results suggest that θ is greater than 0.

We add that for practical purposes the solution above needs to be verified; specifically, several sets of starting values must be examined in order to confirm that the algorithm did not terminate in a local optimum. McCullough (2004) offers further advice on nonlinear estimation.

Note also that the function presented above is specialized to the data set under investigation. If a reusable function is needed, a proper function

GCobbDouglas(formula, data, ...) should be coded and preferably made
available in a package (R Development Core Team 2008g), typically the best
means of sharing collections of functions.

7.4 Reproducible Econometrics Using Sweave()

As noted above, reproducibility of analyses is crucial, both in academic and
corporate environments (Leisch and Rossini 2003). See McCullough and Vinod
(2003), Zeileis and Kleiber (2005), and Koenker and Zeileis (2007) for three
recent examples of partially successful replications in an academic setting. Several features make R an ideal environment for reproducible research. Firstly, R
is mostly platform independent and runs on Windows, the Mac family of operating systems, and various flavors of Unix. Secondly, its open-source nature
not only makes the (binary) system available to everyone but also enables inspection of the full underlying source code if necessary. Moreover, R supports
literate programming: Sweave() (Leisch 2002, 2003a) allows for mixing R and
LaTeX code, making the code and documentation a tightly coupled bundle.

In fact, this book has been entirely written using Sweave() functionality.
For compiling a new version of the book, first the whole source code is executed, its output (text and graphics) is "weaved" with the LaTeX text, and
then pdfLaTeX is run to produce the final book in PDF (portable document
format). Therefore, it is assured that the input and output displayed are always in sync with the versions of the data, code, packages, and R itself. The
whole process is platform independent—incidentally, the authors simultaneously used Microsoft Windows, Mac OS X, and Debian GNU/Linux during
the course of writing this book.

In the following, we illustrate a fairly simple example for mixing R code
and LaTeX documentation. We start out from the file Sweave-journals.Rnw
displayed in Table 7.1.[1] It mainly looks like a LaTeX file, but it contains R code
chunks beginning with <<...>>= and ending in @. This file can be processed
by R upon calling

```
R> Sweave("Sweave-journals.Rnw")
```

which replaces the original R code by valid LaTeX code and weaves it into
the file Sweave-journals.tex shown in Table 7.2. In place of the R chunks,
this contains verbatim LaTeX chunks with the input and output of the R
commands and/or an \includegraphics{} statement for the inclusion of
figures generated along the way. The additional environments, such as Sinput,
Soutput, and Schunk, are defined in the style file Sweave.sty, a file that is
part of the local R installation. It is included automatically with a system-
dependent path. The file Sweave-journals.tex can then be processed as
usual by LaTeX, producing the final document as shown in Table 7.3.

---

[1] The suffix .Rnw is conventionally used to abbreviate "R noweb". Noweb is a simple
literate-programming tool whose syntax is reused in Sweave().

Table 7.1. A simple Sweave file: `Sweave-journals.Rnw`.

```
\documentclass[a4paper]{article}

\begin{document}

We fit a linear regression for the economic journals demand model.

<<>>=
data("Journals", package = "AER")
journals_lm <- lm(log(subs) ~ log(price/citations), data = Journals)
journals_lm
@

A scatter plot with the fitted regression line is shown below.

\begin{center}
<<fig=TRUE, echo=FALSE>>=
plot(log(subs) ~ log(price/citations), data = Journals)
abline(journals_lm)
@
\end{center}

\end{document}
```

In addition to "weaving", there is a second basic operation for literate-programming documents, called "tangling", which here amounts to extracting the included R code. Invoking `Stangle()` via

```
R> Stangle("Sweave-journals.Rnw")
```

produces a file `Sweave-journals.R` that simply contains the R code from the two R chunks.

The basic weaving procedure illustrated above can be refined in many ways. In the starting lines of an R chunk `<<...>>=`, control options can be inserted (together with an optional name for the chunk). Above, we already use `echo=FALSE` (which suppresses the display of the code input) and `fig=TRUE` (which signals that figures should be produced). By default, both EPS (encapsulated PostScript) and PDF files are generated so that the associated LATEX sources can be compiled either with plain LATEX (for DVI documents) or pdfLATEX (for PDF documents). Similarly, many other options can be set, such as height/width for graphics; see `?RweaveLatex` for a full list. For running LATEX, the user can, of course, employ any means he or she is accustomed to. However, if desired, running LATEX is also possible from within R by using

Table 7.2. LaTeX file `Sweave-journals.tex` obtained from `Sweave("Sweave-journals.Rnw")`.

```
\documentclass[a4paper]{article}

\usepackage{/usr/share/R/share/texmf/Sweave}
\begin{document}

We fit a linear regression for the economic journals demand model.

\begin{Schunk}
\begin{Sinput}
R> data("Journals", package = "AER")
R> journals_lm <- lm(log(subs) ~ log(price/citations),
+       data = Journals)
R> journals_lm
\end{Sinput}
\begin{Soutput}
Call:
lm(formula = log(subs) ~ log(price/citations),   data = Journals)

Coefficients:
          (Intercept)   log(price/citations)
               4.766                 -0.533
\end{Soutput}
\end{Schunk}

A scatter plot with the fitted regression line is shown below.

\begin{center}
\includegraphics{Sweave-journals-002}
\end{center}

\end{document}
```

---

the function `texi2dvi()` from the package **tools** [2]. PDF output can then be generated by

```
R> texi2dvi("Sweave-journals.tex", pdf = TRUE)
```

The source code for the example above is also contained in a so-called vignette (Leisch 2003b) in the folder ~/AER/inst/doc of the **AER** package. The associated PDF document can be viewed by calling

---

[2] If `texi2dvi()` does not work out of the box, it might be necessary to set the "texi2dvi" option, see `?texi2dvi` for details.

Table 7.3. Final document created by running LaTeX on `Sweave-journals.tex`.

We fit a linear regression for the economic journals demand model.

```
R> data("Journals", package = "AER")
R> journals_lm <- lm(log(subs) ~ log(price/citations),
+       data = Journals)
R> journals_lm

Call:
lm(formula = log(subs) ~ log(price/citations),   data = Journals)

Coefficients:
          (Intercept)  log(price/citations)
                4.766                -0.533
```

A scatter plot with the fitted regression line is shown below.

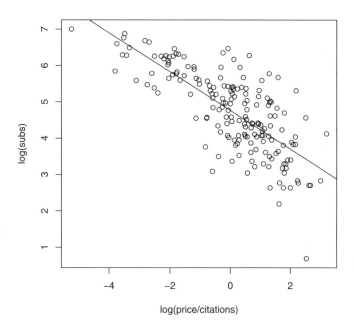

R> vignette("Sweave-journals", package = "AER")

This document makes it obvious that it was generated from R because it includes verbatim R input and output.

For generating reports or papers, one often wants to avoid verbatim sections and use LaTeX formulas and equations instead. This is also easily achieved using `Sweave()`, while maintaining the dynamic character of the document. Two simple ways are to include the output of R expressions directly in the text via `\Sexpr{}` or to produce full LaTeX code in an R chunk. We use the former for displaying a formula for the journals regression along with the estimated coefficients. The LaTeX code for this is

```
\[
  \log(\textrm{subscriptions}) \quad = \quad
    \Sexpr{round(coef(journals_lm)[1], digits = 2)}
    \Sexpr{if(coef(journals_lm)[2] < 0) "-" else "+"}
    \Sexpr{abs(round(coef(journals_lm)[2], digits = 2))}
    \cdot \log(\textrm{price per citation})
\]
```

containing three `\Sexpr{}` statements: the first and third simply extract and round the coefficients of the fitted "lm" model, and the second one dynamically chooses the right sign for the slope. The output in the processed document is

$$\log(\text{subscriptions}) \quad = \quad 4.77 - 0.53 \cdot \log(\text{price per citation})$$

This is very handy for including small text fragments, typically numbers or short sequences of text. For producing more complex text structures (e.g., tables), it is often simpler to use the rich text processing functionality (see Chapter 2 and `?paste`) and put together the full LaTeX code within R.

Here, we produce the usual table of coefficients for a regression model. The code below sets up a table in LaTeX, the core of which is filled by the output of an R chunk whose output is treated as LaTeX code by setting the option `results=tex`. It extracts the coefficients table from the `summary()` of the "lm" object, rounds the numbers and coerces them to characters, prettifies the p values, and then pastes everything into a LaTeX table. The resulting document is shown in Table 7.4.

Table 7.4. Regression summary

	Estimate	Std. error	t statistic	p value
(Intercept)	4.766	0.056	85.249	< 0.001
log(price/citations)	-0.533	0.036	-14.968	< 0.001

```
\begin{table}
\centering
\caption{Regression summary}
\begin{tabular}{rrrrr}
\hline
& Estimate & Std.~error & $t$ statistic & $p$ value \\
\hline
<<echo=FALSE, results=tex>>=
x <- summary(journals_lm)$coefficients
x[] <- as.character(round(x, digits = 3))
x[,4] <- ifelse(as.numeric(x[,4]) < 0.001, "$<$ 0.001", x[,4])
cat(paste(rownames(x), "&",
  apply(x, 1, paste, collapse = " & "), "\\\\ \n"))
@
\hline
\end{tabular}
\end{table}
```

The last command in the R chunk is somewhat more complex; hence we recommend that readers simply try to put it together step by step. Note that the backslash \ has to be escaped because it is the escape character. Therefore, to produce a double backslash \\ in cat(), both backslashes have to be escaped.

Instead of hand-crafting the LaTeX summary, similar code could have been generated using the **xtable** package (Dahl 2007), which can export tables for several standard R objects to LaTeX or HTML.

Although base R just supports weaving of R code and LaTeX documentation, Sweave() is written in such a way that it can be extended to other documentation formats as well. The package **R2HTML** (Lecoutre 2003) provides the function RweaveHTML(), which allows mixing R code with HTML documentation. Similarly, the package **odfWeave** (Kuhn 2008) contains infrastructure for embedding R code and output into word-processor files. Specifically, the function RweaveOdf() and its wrapper odfWeave() can weave R and text in open document format (ODF). This is a word-processor format supported by various programs, most notably the OpenOffice.org suite, from which ODF files can be exported into many other formats, including Microsoft Word documents and rich text format (RTF). See Koenker and Zeileis (2007) for further information.

7.5 Exercises

1. Empirical macroeconomics has seen revived interest in filtering techniques. Write a function implementing the filter known as the Hodrick-Prescott filter among economists, including an estimate of its smoothing parameter. Schlicht (2005) is a good starting point.
2. Our analysis of the OECDGrowth data in Chapter 4 used highly nonlinear robust regression techniques; in addition, the data set is rather small.
   (a) Compute bootstrap standard errors as an alternative to the conventional asymptotic approximations.
   (b) Compare the results in (a) with the standard errors obtained in Chapter 4.
3. Zellner and Ryu (1998) also use the Equipment data while discussing a wider set of functional forms.
   (a) (optional) Do not use the **AER** version of the data set. Instead, download these data from the data archive of the *Journal of Applied Econometrics*, where they are ordered by variable. Use the scan() function to generate a data.frame.
   (b) Estimate some of the generalized functional forms considered by Zellner and Ryu (1998) via maximum likelihood following the approach outlined in Section 7.3.

References

Andrews DWK (1991). "Heteroskedasticity and Autocorrelation Consistent Covariance Matrix Estimation." *Econometrica*, **59**, 817–858.

Andrews DWK (1993). "Tests for Parameter Instability and Structural Change with Unknown Change Point." *Econometrica*, **61**, 821–856.

Andrews DWK, Monahan JC (1992). "An Improved Heteroskedasticity and Autocorrelation Consistent Covariance Matrix Estimator." *Econometrica*, **60**(4), 953–966.

Andrews DWK, Ploberger W (1994). "Optimal Tests When a Nuisance Parameter Is Present Only Under the Alternative." *Econometrica*, **62**, 1383–1414.

Anglin PM, Gençay R (1996). "Semiparametric Estimation of a Hedonic Price Function." *Journal of Applied Econometrics*, **11**, 633–648.

Arellano M, Bond S (1991). "Some Tests of Specification for Panel Data: Monte Carlo Evidence and an Application to Employment Equations." *Review of Economic Studies*, **58**, 277–297.

Bai J, Perron P (1998). "Estimating and Testing Linear Models with Multiple Structural Changes." *Econometrica*, **66**, 47–78.

Bai J, Perron P (2003). "Computation and Analysis of Multiple Structural Change Models." *Journal of Applied Econometrics*, **18**, 1–22.

Balcilar M (2007). *mFilter: Miscellaneous Time-Series Filters*. R package version 0.1-3, URL http://CRAN.R-project.org/package=mFilter.

Baltagi BH (2002). *Econometrics*. 3rd edition. Springer-Verlag, New York. URL http://www.springeronline.com/sgw/cda/frontpage/0, 10735,4-165-2-107420-0,00.html.

Baltagi BH (2005). *Econometric Analysis of Panel Data*. 3rd edition. John Wiley & Sons, Hoboken, NJ. URL http://www.wiley.com/legacy/wileychi/baltagi3e/.

Baltagi BH, Khanti-Akom S (1990). "On Efficient Estimation with Panel Data: An Empirical Comparison of Instrumental Variables Estimators." *Journal of Applied Econometrics*, **5**, 401–406.

Bates D (2008). *lme4: Linear Mixed-Effects Models Using S4 Classes*. R package version 0.99875-9, URL http://CRAN.R-project.org/package=lme4.

Becker RA, Chambers JM (1984). *S: An Interactive Environment for Data Analysis and Graphics*. Wadsworth and Brooks/Cole, Monterey, CA.

Becker RA, Chambers JM, Wilks AR (1988). *The New S Language*. Chapman & Hall, London.

Belsley DA, Kuh E, Welsch RE (1980). *Regression Diagnostics: Identifying Influential Data and Sources of Collinearity*. John Wiley & Sons, New York.

Beran J, Whitcher B, Maechler M (2007). *longmemo: Statistics for Long-Memory Processes*. R package version 0.9-5, URL http://CRAN.R-project.org/package=longmemo.

Bergstrom TC (2001). "Free Labor for Costly Journals?" *Journal of Economic Perspectives*, **15**, 183–198.

Berndt ER (1991). *The Practice of Econometrics*. Addison-Wesley, Reading, MA.

Bierens H, Ginther DK (2001). "Integrated Conditional Moment Testing of Quantile Regression Models." *Empirical Economics*, **26**, 307–324.

Bollerslev T, Ghysels E (1996). "Periodic Autoregressive Conditional Heteroskedasticity." *Journal of Business & Economic Statistics*, **14**, 139–151.

Box GEP, Jenkins G (1970). *Time Series Analysis: Forecasting and Control*. Holden-Day, San Francisco.

Box GEP, Pierce DA (1970). "Distribution of Residual Correlations in Autoregressive-Integrated Moving Average Time Series Models." *Journal of the American Statistical Association*, **65**, 1509–1526.

Breusch TS (1979). "Testing for Autocorrelation in Dynamic Linear Models." *Australian Economic Papers*, **17**, 334–355.

Breusch TS, Pagan AR (1979). "A Simple Test for Heteroscedasticity and Random Coefficient Variation." *Econometrica*, **47**, 1287–1294.

Brockwell PJ, Davis RA (1991). *Time Series: Theory and Methods*. 2nd edition. Springer-Verlag, New York.

Brockwell PJ, Davis RA (1996). *Introduction to Time Series and Forecasting*. Springer-Verlag, New York.

Brown RL, Durbin J, Evans JM (1975). "Techniques for Testing the Constancy of Regression Relationships over Time." *Journal of the Royal Statistical Society, Series B*, **37**, 149–163.

Cameron AC, Trivedi PK (1990). "Regression-Based Tests for Overdispersion in the Poisson Model." *Journal of Econometrics*, **46**, 347–364.

Cameron AC, Trivedi PK (1998). *Regression Analysis of Count Data*. Cambridge University Press, Cambridge.

Canty A, Ripley BD (2008). *boot: Functions and Datasets for Bootstrapping*. R package version 1.2-33, URL http://CRAN.R-project.org/package=boot.

Chambers JM (1998). *Programming with Data*. Springer-Verlag, New York.

Chambers JM, Hastie TJ (eds.) (1992). *Statistical Models in S*. Chapman & Hall, London.

Cleveland RB, Cleveland WS, McRae J, Terpenning I (1990). "STL: A Seasonal-Trend Decomposition Procedure Based on Loess." *Journal of Official Statistics*, **6**, 3–73.

Cleveland WS (1993). *Visualizing Data*. Hobart Press, Summit, NJ.

Cornwell C, Rupert P (1988). "Efficient Estimation with Panel Data: An Empirical Comparison of Instrumental Variables Estimators." *Journal of Applied Econometrics*, **3**, 149–155.

Cribari-Neto F (2004). "Asymptotic Inference Under Heteroskedasticity of Unknown Form." *Computational Statistics & Data Analysis*, **45**, 215–233.

Cribari-Neto F, Zarkos SG (1999). "R: Yet Another Econometric Programming Environment." *Journal of Applied Econometrics*, **14**, 319–329.

Croissant Y (2008). *mlogit: Multinomial Logit Models with Choice-Specific Variables*. R package version 0.1-0, URL http://CRAN.R-project.org/package=mlogit.

Croissant Y, Millo G (2008). "**plm**: Linear Models for Panel Data." *Journal of Statistical Software*. Forthcoming.

Dahl DB (2007). *xtable: Export Tables to LaTeX or HTML*. R package version 1.5-2, URL http://CRAN.R-project.org/package=xtable.

Dalgaard P (2002). *Introductory Statistics with R*. Springer-Verlag, New York.

Davidson R, MacKinnon JG (1981). "Several Tests for Model Specification in the Presence of Alternative Hypotheses." *Econometrica*, **49**, 781–793.

Davidson R, MacKinnon JG (2004). *Econometric Theory and Methods*. Oxford University Press, Oxford.

Davison AC, Hinkley DV (1997). *Bootstrap Methods and Their Applications*. Cambridge University Press, Cambridge.

Di Narzo AF, Aznarte JL (2008). *tsDyn: Time Series Analysis Based on Dynamical Systems Theory*. R package version 0.6-0, URL http://CRAN.R-project.org/package=tsDyn.

Dickey D, Fuller W (1981). "Likelihood Ratio Tests for Autoregressive Time Series with a Unit Root." *Econometrica*, **49**, 1057–1072.

Durbin J, Koopman SJ (2001). *Time Series Analysis by State Space Methods*. Oxford University Press, Oxford.

Durbin J, Watson GS (1950). "Testing for Serial Correlation in Least Squares Regression I." *Biometrika*, **37**, 409–428.

Eicker F (1963). "Asymptotic Normality and Consistency of the Least Squares Estimator for Families of Linear Regressions." *Annals of Mathematical Statistics*, **34**, 447–456.

Elliott G, Rothenberg TJ, Stock JH (1996). "Efficient Tests for an Autoregressive Unit Root." *Econometrica*, **64**, 813–836.

Engle RF, Granger CWJ (1987). "Co-integration and Error Correction: Representation, Estimation, and Testing." *Econometrica*, **55**, 251–276.

Fair RC (1978). "A Theory of Extramarital Affairs." *Journal of Political Economy*, **86**, 45–61.

Faraway JJ (2005). *Linear Models with R*. Chapman & Hall/CRC, Boca Raton, FL.

Fox J (2002). *An R and S-PLUS Companion to Applied Regression*. Sage Publications, Thousand Oaks, CA.

Fox J (2003). "Effect Displays in R for Generalised Linear Models." *Journal of Statistical Software*, **8**(15), 1–18. URL http://www.jstatsoft.org/v8/i15/.

Fraley C, Leisch F, Maechler M (2006). **fracdiff**: *Fractionally Differenced ARIMA Models*. R package version 1.3-1, URL http://CRAN.R-project.org/package=fracdiff.

Franses PH (1998). *Time Series Models for Business and Economic Forecasting*. Cambridge University Press, Cambridge. URL http://www.few.eur.nl/few/people/franses/research/book2.htm.

Freedman DA (2006). "On the So-Called 'Huber Sandwich Estimator' and 'Robust Standard Errors'." *The American Statistician*, **60**(4), 299–302.

Friendly M (1994). "Mosaic Displays for Multi-Way Contingency Tables." *Journal of the American Statistical Association*, **89**, 190–200.

Gerfin M (1996). "Parametric and Semi-Parametric Estimation of the Binary Response Model of Labour Market Participation." *Journal of Applied Econometrics*, **11**, 321–339.

Gilbert P (2007). **dse**: *Dynamic Systems Estimation*. R bundle with packages **dse1** and **dse2**, version 2007.11-1, URL http://CRAN.R-project.org/package=dse.

Godfrey LG (1978). "Testing Against General Autoregressive and Moving Average Error Models when the Regressors Include Lagged Dependent Variables." *Econometrica*, **46**, 1293–1302.

Goldberg D (1991). "What Every Computer Scientist Should Know About Floating-Point Arithmetic." *ACM Computing Surveys*, **23**, 5–48.

Goldfeld SM, Quandt RE (1965). "Some Tests for Homoskedasticity." *Journal of the American Statistical Association*, **60**, 539–547.

Graves S (2008). **FinTS**: *Companion to Tsay's 'Analysis of Financial Time Series'*. R package version 0.3-1, URL http://CRAN.R-project.org/package=FinTS.

Greene WH (1993). *Econometric Analysis*. 2nd edition. Macmillan Publishing Company, New York.

Greene WH (2003). *Econometric Analysis*. 5th edition. Prentice Hall, Upper Saddle River, NJ. URL http://pages.stern.nyu.edu/~wgreene/Text/econometricanalysis.htm.

Grothendieck G (2005). **dyn**: *Time Series Regression*. R package version 0.2-6, URL http://CRAN.R-project.org/package=dyn.

Grothendieck G, Petzoldt T (2004). "R Help Desk: Date and Time Classes in R." *R News*, **4**(1), 29–32. URL http://CRAN.R-project.org/doc/Rnews/.

Grunfeld Y (1958). "The Determinants of Corporate Investment." Unpublished Ph.D. dissertation, Department of Economics, University of Chicago.

Gurmu S, Trivedi PK (1996). "Excess Zeros in Count Models for Recreational Trips." *Journal of Business & Economic Statistics*, **14**, 469–477.

Hamilton JD (1994). *Time Series Analysis*. Princeton University Press, Princeton, NJ.

Hartigan JA, Kleiner B (1981). "Mosaics for Contingency Tables." In W Eddy (ed.), "Computer Science and Statistics: Proceedings of the 13th Symposium on the Interface," pp. 268–273. Springer-Verlag, New York.

Harvey AC (1989). *Forecasting, Structural Time Series Models and the Kalman Filter*. Cambridge University Press, Cambridge.

Harvey AC, Collier P (1977). "Testing for Functional Misspecification in Regression Analysis." *Journal of Econometrics*, **6**, 103–119.

Harvey AC, Durbin J (1986). "The Effects of Seat Belt Legislation on British Road Casualties: A Case Study in Structural Time Series Modelling." *Journal of the Royal Statistical Society, Series A*, **149**, 187–227. With discussion.

Hastie T (2006). *gam: Generalized Additive Models*. R package version 0.98, URL http://CRAN.R-project.org/package=gam.

Hastie T, Tibshirani R, Friedman J (2001). *The Elements of Statistical Learning*. Springer-Verlag, New York.

Hausman JA, Taylor WE (1981). "Panel Data and Unobservable Individual Effects." *Econometrica*, **49**, 1377–1398.

Hayfield T, Racine JS (2008). "Nonparametric Econometrics: The **np** Package." *Journal of Statistical Software*. Forthcoming.

Heij C, de Boer PMC, Franses PH, Kloek T, van Dijk HK (2004). *Econometric Methods with Applications in Business and Economics*. Oxford University Press, Oxford.

Henningsen A (2008). "Demand Analysis with the Almost Ideal Demand System in R: Package **micEcon**." *Journal of Statistical Software*. Forthcoming.

Henningsen A, Hamann JD (2007). "**systemfit**: A Package for Estimating Systems of Simultaneous Equations in R." *Journal of Statistical Software*, **23**(4), 1–40. URL http://www.jstatsoft.org/v23/i04/.

Honda Y (1985). "Testing the Error Components Model with Non-normal Disturbances." *Review of Economic Studies*, **52**, 681–690.

Hyndman RJ, Khandakar Y (2008). "Automatic Time-Series Forecasting: The **forecast** Package for R." *Journal of Statistical Software*. Forthcoming.

Ihaka R, Gentleman R (1996). "R: A Language for Data Analysis and Graphics." *Journal of Computational and Graphical Statistics*, **5**, 299–314.

Johansen S (1991). "Estimation and Hypothesis Testing of Cointegration Vectors in Gaussian Vector Autoregressive Models." *Econometrica*, **59**, 1551–1580.

Klein RW, Spady RH (1993). "An Efficient Semiparametric Estimator for Binary Response Models." *Econometrica*, **61**, 387–421.

Koenker R (1981). "A Note on Studentizing a Test for Heteroscedasticity." *Journal of Econometrics*, **17**, 107–112.

Koenker R (2005). *Quantile Regression*. Cambridge University Press, Cambridge.

Koenker R (2008). *quantreg: Quantile Regression*. R package version 4.17, URL http://CRAN.R-project.org/package=quantreg.

Koenker R, Hallock KF (2001). "Quantile Regression." *Journal of Economic Perspectives*, **15**(4), 143–156.

Koenker R, Zeileis A (2007). "Reproducible Econometric Research (A Critical Review of the State of the Art)." *Report 60*, Department of Statistics and Mathematics, Wirtschaftsuniversität Wien, Research Report Series. URL http://epub.wu-wien.ac.at/.

Krämer W, Sonnberger H (1986). *The Linear Regression Model Under Test.* Physica-Verlag, Heidelberg.

Kuhn M (2008). *odfWeave:* Sweave *Processing of Open Document Format (ODF) Files.* R package version 0.7.5, URL http://CRAN.R-project.org/package=odfWeave.

Kwiatkowski D, Phillips PCB, Schmidt P, Shin Y (1992). "Testing the Null Hypothesis of Stationarity against the Alternative of a Unit Root." *Journal of Econometrics*, **54**, 159–178.

Lambert D (1992). "Zero-Inflated Poisson Regression, with an Application to Defects in Manufacturing." *Technometrics*, **34**, 1–14.

Lecoutre E (2003). "The **R2HTML** Package." *R News*, **3**(3), 33–36. URL http://CRAN.R-project.org/doc/Rnews/.

Leisch F (2002). "Dynamic Generation of Statistical Reports Using Literate Data Analysis." In W Härdle, B Rönz (eds.), "COMPSTAT 2002 – Proceedings in Computational Statistics," pp. 575–580. Physica Verlag, Heidelberg.

Leisch F (2003a). "Sweave and Beyond: Computations on Text Documents." In K Hornik, F Leisch, A Zeileis (eds.), "Proceedings of the 3rd International Workshop on Distributed Statistical Computing, Vienna, Austria," ISSN 1609-395X, URL http://www.ci.tuwien.ac.at/Conferences/DSC-2003/Proceedings/.

Leisch F (2003b). "Sweave, Part II: Package Vignettes." *R News*, **3**(2), 21–24. URL http://CRAN.R-project.org/doc/Rnews/.

Leisch F, Rossini AJ (2003). "Reproducible Statistical Research." *Chance*, **16**(2), 46–50.

Li Q, Racine JS (2007). *Nonparametric Econometrics. Theory and Practice.* Princeton University Press, Princeton, NJ.

Ligges U (2007). *Programmieren in R.* 2nd edition. Springer-Verlag, Berlin.

Ljung GM, Box GEP (1978). "On a Measure of Lack of Fit in Time Series Models." *Biometrika*, **65**, 553–564.

Long JS, Ervin LH (2000). "Using Heteroscedasticity Consistent Standard Errors in the Linear Regression Model." *The American Statistician*, **54**, 217–224.

Lumley T, Heagerty P (1999). "Weighted Empirical Adaptive Variance Estimators for Correlated Data Regression." *Journal of the Royal Statistical Society, Series B*, **61**, 459–477.

Lütkepohl H, Teräsvirta T, Wolters J (1999). "Investigating Stability and Linearity of a German M1 Money Demand Function." *Journal of Applied Econometrics*, **14**, 511–525.

MacKinnon JG, White H (1985). "Some Heteroskedasticity-Consistent Covariance Matrix Estimators with Improved Finite Sample Properties." *Journal of Econometrics*, **29**, 305–325.

Maddala GS (2001). *Introduction to Econometrics*. 3rd edition. John Wiley & Sons, New York.

Maechler M, Rousseeuw P, Croux C, Todorov V, Ruckstuhl A, Salibian-Barrera M (2007). *robustbase: Basic Robust Statistics*. R package version 0.2-8, URL http://CRAN.R-project.org/package=robustbase.

Mankiw NG, Romer D, Weil DN (1992). "A Contribution to the Empirics of Economic Growth." *Quarterly Journal of Economics*, **107**, 407–437.

McCullagh P, Nelder JA (1989). *Generalized Linear Models*. 2nd edition. Chapman & Hall, London.

McCullough BD (2004). "Some Details of Nonlinear Estimation." In M Altman, J Gill, MP McDonald (eds.), "Numerical Issues in Statistical Computing for the Social Scientist," pp. 199–218. John Wiley & Sons, Hoboken, NJ.

McCullough BD, Vinod HD (2003). "Verifying the Solution from a Nonlinear Solver: A Case Study." *American Economic Review*, **93**, 873–892.

Meyer D (2002). "Naive Time Series Forecasting Methods." *R News*, **2**(2), 7–10. URL http://CRAN.R-project.org/doc/Rnews/.

Mroz TA (1987). "The Sensitivity of an Empirical Model of Married Women's Hours of Work to Economic and Statistical Assumptions." *Econometrica*, **55**, 765–799.

Mullahy J (1986). "Specification and Testing of Some Modified Count Data Models." *Journal of Econometrics*, **33**, 341–365.

Murrell P (2005). *R Graphics*. Chapman & Hall/CRC, Boca Raton, FL.

Murrell P, Ihaka R (2000). "An Approach to Providing Mathematical Annotation in Plots." *Journal of Computational and Graphical Statistics*, **9**(3), 582–599.

Nelder JA, Wedderburn RWM (1972). "Generalized Linear Models." *Journal of the Royal Statistical Society, Series A*, **135**, 370–384.

Nelson CR, Plosser CI (1982). "Trends and Random Walks in Economic Time Series: Some Evidence and Implications." *Journal of Monetary Economics*, **10**, 139–162.

Newey WK, West KD (1987). "A Simple, Positive-Definite, Heteroskedasticity and Autocorrelation Consistent Covariance Matrix." *Econometrica*, **55**, 703–708.

Newey WK, West KD (1994). "Automatic Lag Selection in Covariance Matrix Estimation." *Review of Economic Studies*, **61**, 631–653.

Nonneman W, Vanhoudt P (1996). "A Further Augmentation of the Solow Model and the Empirics of Economic Growth for OECD Countries." *Quarterly Journal of Economics*, **111**, 943–953.

Ozuna T, Gomez I (1995). "Specification and Testing of Count Data Recreation Demand Functions." *Empirical Economics*, **20**, 543–550.

Pfaff B (2006). *Analysis of Integrated and Cointegrated Time Series with R.* Springer-Verlag, New York.

Pfaff B (2008). "VAR, SVAR and SVEC Models: Implementation Within R Package **vars**." *Journal of Statistical Software.* Forthcoming.

Phillips PCB, Ouliaris S (1990). "Asymptotic Properties of Residual Based Tests for Cointegration." *Econometrica,* **58**, 165–193.

Phillips PCB, Perron P (1988). "Trends and Random Walks in Macroeconomic Time Series." *Biometrika,* **75**, 335–346.

Ploberger W, Krämer W (1992). "The CUSUM Test with OLS Residuals." *Econometrica,* **60**(2), 271–285.

Ploberger W, Krämer W, Kontrus K (1989). "A New Test for Structural Stability in the Linear Regression Model." *Journal of Econometrics,* **40**, 307–318.

Racine J, Hyndman RJ (2002). "Using R to Teach Econometrics." *Journal of Applied Econometrics,* **17**, 175–189.

Ramsey JB (1969). "Tests for Specification Error in Classical Linear Least Squares Regression Analysis." *Journal of the Royal Statistical Society, Series B,* **31**, 350–371.

R Development Core Team (2008a). *An Introduction to R.* R Foundation for Statistical Computing, Vienna, Austria. ISBN 3-900051-12-7, URL http://www.R-project.org/.

R Development Core Team (2008b). *R: A Language and Environment for Statistical Computing.* R Foundation for Statistical Computing, Vienna, Austria. ISBN 3-900051-07-0, URL http://www.R-project.org/.

R Development Core Team (2008c). *R Data Import/Export.* R Foundation for Statistical Computing, Vienna, Austria. ISBN 3-900051-10-0, URL http://www.R-project.org/.

R Development Core Team (2008d). *R Installation and Administration.* R Foundation for Statistical Computing, Vienna, Austria. ISBN 3-900051-09-7, URL http://www.R-project.org/.

R Development Core Team (2008e). *R Internals.* R Foundation for Statistical Computing, Vienna, Austria. ISBN 3-900051-14-3, URL http://www.R-project.org/.

R Development Core Team (2008f). *R Language Definition.* R Foundation for Statistical Computing, Vienna, Austria. ISBN 3-900051-13-5, URL http://www.R-project.org/.

R Development Core Team (2008g). *Writing R Extensions.* R Foundation for Statistical Computing, Vienna, Austria. ISBN 3-900051-11-9, URL http://www.R-project.org/.

Ripley BD (2002). "Time Series in R 1.5.0." *R News,* **2**(2), 2–7. URL http://CRAN.R-project.org/doc/Rnews/.

Rousseeuw PJ (1984). "Least Median of Squares Regression." *Journal of the American Statistical Association,* **79**, 871–880.

Sarkar D (2002). "**lattice**: An Implementation of Trellis Graphics in R." *R News,* **2**(2), 19–23. URL http://CRAN.R-project.org/doc/Rnews/.

Schlicht E (2005). "Estimating the Smoothing Parameter in the So-Called Hodrick-Prescott Filter." *Journal of the Japan Statistical Society*, **35**, 99–119.

Sing T, Sander O, Beerenwinkel N, Lengauer T (2005). "**ROCR**: Visualizing Classifier Performance in R." *Bioinformatics*, **21**(20), 3940–3941.

Steel GL, Sussman GJ (1975). "Scheme: An Interpreter for the Extended Lambda Calculus." *Memo 349*, MIT Artificial Intelligence Laboratory.

Stock JH, Watson MW (2007). *Introduction to Econometrics*. 2nd edition. Addison-Wesley, Reading, MA.

Stokes H (2004). "On the Advantage of Using Two or More Econometric Software Systems to Solve the Same Problem." *Journal of Economic and Social Measurement*, **29**, 307–320.

Therneau TM, Grambsch PM (2000). *Modeling Survival Data*. Springer-Verlag, New York.

Tobin J (1958). "Estimation of Relationships for Limited Dependent Variables." *Econometrica*, **26**, 24–36.

Toomet O, Henningsen A (2008). "Sample Selection Models in R: Package **sampleSelection**." *Journal of Statistical Software*. Forthcoming.

Trapletti A (2008). ***tseries: Time Series Analysis and Computational Finance***. R package version 0.10-15, URL http://CRAN.R-project.org/package=tseries.

Tsay RS (2005). *Analysis of Financial Time Series*. 2nd edition. John Wiley & Sons, Hoboken, NJ.

Utts JM (1982). "The Rainbow Test for Lack of Fit in Regression." *Communications in Statistics – Theory and Methods*, **11**, 1801–1815.

Venables WN, Ripley BD (2000). *S Programming*. Springer-Verlag, New York.

Venables WN, Ripley BD (2002). *Modern Applied Statistics with S*. 4th edition. Springer-Verlag, New York.

Verbeek M (2004). *A Guide to Modern Econometrics*. 2nd edition. John Wiley & Sons, Hoboken, NJ.

Vinod HD (2006). "Maximum Entropy Ensembles for Time Series Inference in Economics." *Journal of Asian Economics*, **17**(6), 955–978.

White H (1980). "A Heteroskedasticity-Consistent Covariance Matrix Estimator and a Direct Test for Heteroskedasticity." *Econometrica*, **48**, 817–838.

Wilkinson GN, Rogers CE (1973). "Symbolic Description of Factorial Models for Analysis of Variance." *Applied Statistics*, **22**, 392–399.

Wood SN (2006). *Generalized Additive Models: An Introduction with R*. Chapman & Hall/CRC, Boca Raton, FL.

Wooldridge JM (2002). *Econometric Analysis of Cross-Section and Panel Data*. MIT Press, Cambridge, MA.

Wuertz D (2008). ***Rmetrics: An Environment and Software Collection for Teaching Financial Engineering and Computational Finance***. R packages **fArma**, **fAsianOptions**, **fAssets**, **fBasics**, **fCalendar**, **fCopulae**, **fEcofin**, **fExoticOptions**, **fExtremes**, **fGarch**, **fImport**, **fMultivar**, **fNonlinear**, **fOptions**,

fPortfolio, fRegression, fSeries, fTrading, fUnitRoots, fUtilities, URL `http://CRAN.R-project.org/package=Rmetrics`.

Zaman A, Rousseeuw PJ, Orhan M (2001). "Econometric Applications of High-Breakdown Robust Regression Techniques." *Economics Letters*, **71**, 1–8.

Zeileis A (2004). "Econometric Computing with HC and HAC Covariance Matrix Estimators." *Journal of Statistical Software*, **11**(10), 1–17. URL `http://www.jstatsoft.org/v11/i10/`.

Zeileis A (2005). "A Unified Approach to Structural Change Tests Based on ML Scores, *F* Statistics, and OLS Residuals." *Econometric Reviews*, **24**(4), 445–466.

Zeileis A (2006a). "Implementing a Class of Structural Change Tests: An Econometric Computing Approach." *Computational Statistics & Data Analysis*, **50**, 2987–3008.

Zeileis A (2006b). "Object-oriented Computation of Sandwich Estimators." *Journal of Statistical Software*, **16**(9), 1–16. URL `http://www.jstatsoft.org/v16/i09/`.

Zeileis A (2008). *dynlm: Dynamic Linear Regression*. R package version 0.2-0, URL `http://CRAN.R-project.org/package=dynlm`.

Zeileis A, Grothendieck G (2005). "**zoo**: S3 Infrastructure for Regular and Irregular Time Series." *Journal of Statistical Software*, **14**(6), 1–27. URL `http://www.jstatsoft.org/v14/i06/`.

Zeileis A, Hothorn T (2002). "Diagnostic Checking in Regression Relationships." *R News*, **2**(3), 7–10. URL `http://CRAN.R-project.org/doc/Rnews/`.

Zeileis A, Kleiber C (2005). "Validating Multiple Structural Change Models – A Case Study." *Journal of Applied Econometrics*, **20**, 685–690.

Zeileis A, Kleiber C, Jackman S (2008). "Regression Models for Count Data in R." *Journal of Statistical Software*. Forthcoming.

Zeileis A, Kleiber C, Krämer W, Hornik K (2003). "Testing and Dating of Structural Changes in Practice." *Computational Statistics & Data Analysis*, **44**(1–2), 109–123.

Zeileis A, Leisch F, Hornik K, Kleiber C (2002). "**strucchange**: An R Package for Testing for Structural Change in Linear Regression Models." *Journal of Statistical Software*, **7**(2), 1–38. URL `http://www.jstatsoft.org/v07/i02/`.

Zeileis A, Leisch F, Kleiber C, Hornik K (2005). "Monitoring Structural Change in Dynamic Econometric Models." *Journal of Applied Econometrics*, **20**, 99–121.

Zellner A (1962). "An Efficient Method of Estimating Seemingly Unrelated Regressions and Tests for Aggregation Bias." *Journal of the American Statistical Association*, **57**, 348–368.

Zellner A, Revankar N (1969). "Generalized Production Functions." *Review of Economic Studies*, **36**, 241–250.

Zellner A, Ryu H (1998). "Alternative Functional Forms for Production, Cost and Returns to Scale Functions." *Journal of Applied Econometrics*, **13**, 101–127.

Index

springer.com

Introductory Statistics with R
Second Edition

Peter Dalgaard

This book provides an elementary-level introduction to R, targeting both non-statistician scientists in various fields and students of statistics. The main mode of presentation is via code examples with liberal commenting of the code and the output, from the computational as well as the statistical viewpoint. A supplementary R package can be downloaded and contains the data sets. In the second edition, the text and code have been updated to R version 2.6.2. The last two methodological chapters are new, as is a chapter on advanced data handling. The introductory chapter has been extended and reorganized as two chapters. Exercises have been revised and answers are now provided in an Appendix.

2008. Approx. 384 pp. (Statistics and Computing) Softcover
ISBN 978-0-387-79053-4

Data Manipulation with R

Phil Spector

This book presents a wide array of methods applicable for reading data into R, and efficiently manipulating that data. In addition to the built-in functions, a number of readily available packages from CRAN (the Comprehensive R Archive Network) are also covered. All of the methods presented take advantage of the core features of R: vectorization, efficient use of subscripting, and the proper use of the varied functions in R that are provided for common data management tasks.

2008. 164 pp. (Use R!) Softcover
ISBN 978-0-387-74730-9

Statistical Methods for Financial Markets

Tze Leung Lai and Haipeng Xing

This book presents statistical methods and models of importance to quantitative finance and links finance theory to market practice via statistical modeling and decision making. Part I provides basic background in statistics, which includes linear regression and extensions to generalized linear models and nonlinear regression, multivariate analysis, likelihood inference and Bayesian methods, and time series analysis. Part II presents advanced topics in quantitative finance and introduces a substantive-empirical modeling approach to address the discrepancy between finance theory and market data.

2008. 354 pp. (Springer Texts in Statistics) Hardcover
ISBN 978-0-387-77826-6